UNDERSTANDING DYNAMICS
OF
GEOGRAPHIC DOMAINS

UNIVERSITY CONSORTIUM FOR
GEOGRAPHIC INFORMATION SCIENCE

UNDERSTANDING DYNAMICS OF GEOGRAPHIC DOMAINS

EDITED BY

KATHLEEN STEWART HORNSBY
MAY YUAN

CRC Press
Taylor & Francis Group
Boca Raton London New York

CRC Press is an imprint of the
Taylor & Francis Group, an **informa** business

CRC Press
Taylor & Francis Group
6000 Broken Sound Parkway NW, Suite 300
Boca Raton, FL 33487-2742

First issued in paperback 2019

ISBN-13: 978-1-4200-6034-8 (hbk)
ISBN-13: 978-0-367-38752-5 (pbk)

Library of Congress Cataloging-in-Publication Data

Hornsby, Kathleen Stewart
 Understanding dynamics of geographic domains / editors, Kathleen Stewart Hornsby and May Yuan.
     p. cm.
 Includes bibliographical references and index.
 ISBN 978-1-4200-6034-8 (acid-free paper) 1. Geography--Data processing.
 2. Geography--Computer simulation. 3. Spatial data infrastructures. 4.
 Geography--Mathematics. 5. Dynamics. I. Yuan, May. II. Title.

 G70.2.Y83 2008
 910.285--dc22

 2007043271

Visit the Taylor & Francis Web site at
http://www.taylorandfrancis.com

and the CRC Press Web site at
http://www.crcpress.com

Contents

PART I Cognitive Aspects, Representation, and Data Models

Chapter 1

Colin Ware

Chapter 2

Michael F. Goodchild and Alan Glennon

Chapter 3

Steven D. Prager

PART II Analysis, Computation, and Modeling

Chapter 4

Daniel W. Goldberg, John P. Wilson, and Craig A. Knoblock

Chapter 5

Jaymie R. Meliker

PART 3   Visualization and Simulation

Preface

FOCUSING ON DYNAMICS IN GEOGRAPHIC DOMAINS

The world is a dynamic place where many different kinds of happenings take place, such as a crowd gathering in a city square, tornadoes moving across a region, or a plane taking off from an airport. These happenings may have already occurred, may be occurring at the current time, or may be future events for which we want to plan, administer, or forecast. They may be occurrences with which we are very familiar or have never experienced previously. Although this dynamic aspect of the world is widely recognized, information systems have lagged in their ability to represent these dynamics and provide support for users and analysts who work with dynamic domains. The most common approach to data modeling, reinforced through automated imagery data collection, has been the view of the world as a succession of temporal snapshots. These snapshots refer to the representation of a geographic domain at a single moment in time. Recognition of the increased need for richer temporal support that goes beyond a fixed snapshot, combined with a better understanding of dynamics, has spurred the geographic information science research community to focus on the topic of dynamics and time-varying phenomena. This research has introduced approaches for how to incorporate time in geographic information systems (Langran 1992; Worboys 1994; Peuquet 1994; Yuan 1996; Yuan 1999; Peuquet 2002), including different kinds of time (Frank 1998), such as cycles (Hornsby et al. 1999; Campos and Hornsby 2004). Researchers concerned with dynamics have proposed models for *change*, another topic central to dynamics. For example, models have been presented for identity changes that capture how geographic objects evolve over time (Hornsby and Egenhofer 1997; Hornsby and Egenhofer 2000; Worboys 2001; Galton 2003; Galton 2005; Sriti et al. 2005) and change based on spatial and nonspatial attribute alterations (Marceau et al. 2001; Chen and Jiang 2000; Galton 2004; Egenhofer and Wilmsen 2006; Goodchild et al. 2007). Many different application areas highlight the need for understanding dynamics and change — for example, urban development, where a better understanding of dynamics as it relates to urban morphology and land use supports improved urban planning and growth (Parker et al. 2003; Batty 2005; Huang and Claramunt 2005).

In addition to change, a perspective that incorporates happenings or *events* (e.g., traffic jams and volcanic eruptions) also offers opportunities for representing, analyzing, and comprehending dynamics in geographic domains. Recent studies have shown that, in addition to snapshot approaches, the sole use of object-oriented perspectives for representing dynamics is often too restrictive and lacking with respect to capturing semantics of active domains. Since these events share strong links or relations with objects as well as other events, an event-based focus offers a way forward for revealing the rich semantics and characteristics of dynamic entities (Peuquet and Duan 1995; Yuan 2001; Grenon and Smith 2004; Worboys and Hornsby 2004;

Worboys 2005; Galton 2005; McIntosh and Yuan 2005; Galton and Worboys 2005; Cole and Hornsby 2005; Goodchild et al. 2007).

The study of geographic dynamics frequently focuses on movement — for example, configurations of individual or group movements that give us insights into categories of motion including directional movements, and converging or diverging movements (Laube et al. 2005; Stewart Hornsby and Cole 2007). Studies of movement also describe human activity patterns (Kwan 2001; Kwan 2004; Miller 2005) where the temporal patterns of activities are tracked, as well as nonhuman forms of movement where, for example, animal movements are studied (Bennett and Tang 2006), affording opportunities for better planning and management of resources.

These topics are some of the research areas that contribute to the study of dynamics in geographic domains. In this book, we showcase *additional research* that is currently being undertaken in this field. This includes new perspectives on theoretical aspects of geographic dynamics as well as novel computational modeling and visualization-related studies. This book stems from a workshop held in October 2006 on "Computation and Visualization for the Understanding of Dynamics in Geographic Domains," where researchers from academia and members of the U.S. Intelligence Community gathered to discuss dynamics in geographic domains and to learn from each other about the challenges of working with dynamics and developing information systems that are responsive and provide support for dynamics. The following sections give more details about both the workshop and the content of this book.

UCGIS WORKSHOP ON THE COMPUTATION AND VISUALIZATION FOR THE UNDERSTANDING OF DYNAMICS IN GEOGRAPHIC DOMAINS

May Yuan (University of Oklahoma) and Kathleen Stewart Hornsby (University of Iowa) co-organized the workshop "Computation and Visualization for the Understanding of Dynamics in Geographic Domains." The workshop was held in October 2006 at the Maritime Institute of Technology and Graduate Studies, Linthicum Heights, Maryland, under the auspices of the University Consortium for Geographic Information Science (UCGIS) with sponsorship from the Disruptive Technology Office (DTO) on behalf of the U.S. Intelligence Community. A full report of this workshop including the research agenda generated from the workshop is available from CRC Press (Yuan and Stewart Hornsby 2007). For additional information, see also the workshop Web site (http://www.ucgis.org/dynamics_workshop/).

The UCGIS is an organization of universities and professional associations in the United States that have come together to:

- Serve as an effective, unified voice for the geographic information science research community
- Foster multidisciplinary research and education
- Promote the informed and responsible use of geographic information science and geographic analysis for the benefit of society (http://www.ucgis.org/)

The Intelligence Community–UCGIS collaboration began with an earlier work-shop on "Geospatial Visualization and Knowledge Discovery," held in November 2003 in Lansdowne, Virginia (http://www.ucgis.org/Visualization/). This workshop focused on assessing the capabilities of geospatial-visualization and knowledge-discovery techniques for understanding the global security environment based on the extraction, filtering, synthesis, and communication of geospatial intelligence infor-mation. The workshop had a particular focus on detecting unusual events in space and time, and application areas included terrorism, homeland security, spatial epide-miology, and environmental change detection, among others.

The success of this joint workshop laid the groundwork for future efforts, and planning began for the next shared workshop between the Intelligence Community and the UCGIS. In April 2006, a call for position papers was distributed to all UCGIS member institutions, including universities, professional organizations, corporate affiliates, U.S. government agencies, and two international associations. Members of the workshop-steering committee reviewed position papers, and, based on these reviews, 14 geographic information scientists were invited to attend the workshop in October 2006. Additional UCGIS participants, including representatives from industry, brought the number of nongovernment attendees to 25. Rounding out the participants were a further 25 representatives from the U.S. Intelligence Community. In addition, four leading academics working on topics relating to dynamics in geo-graphic domains were invited to deliver plenary talks at the workshop.

The workshop offered a multidisciplinary and multi-agency perspective on the latest research work relating to visualization and computation of geographic dynamics. UCGIS participants represented geography, spatial information science and engineering, history, epidemiology, environmental science, computer science, and cognitive science, among other disciplines. One workshop participant was a representative from Geomatics for Informed Decisions (GEOIDE) in Canada, an organization similar to the UCGIS. Government participants were from the U.S. Intelligence Community, the United States Geological Survey (USGS), and the National Institutes of Health (NIH).

The workshop was organized into sessions of plenary presentations, scenario analysis, government panel discussions, academic-research briefings, and break-out group discussions. In the plenary session, four speakers addressed topics that included visual analysis of urban terrain dynamics; visual analysis of human activi-ties; reasoning about dynamic geospatial data from the perspective of human visual cognition; and a framework for the representation and computation of geographic dynamics. These initial presentations set the stage for provoking the many discus-sions that followed in subsequent workshop sessions. Following the plenary session, a problem-solving exercise gave workshop participants hands-on experience with the nature of intelligence reporting as well as evidence-based intelligence analysis. Participants worked in groups to generate hypotheses about possible events in space and time and construct arguments in support of these hypotheses. Throughout the exercise, the participants discussed and considered visual and computational meth-ods, their application, and their effectiveness for spatiotemporal analysis and predic-tion of events.

A panel of representatives from the U.S. Intelligence Community outlined the needs and challenges of spatiotemporal analysis and modeling for their agencies. Supporting these needs and challenges in tandem with an understanding of activities, events, and processes in geographic domains will help produce intelligence that can be acted upon and that is relevant to achieving their missions. In response, UCGIS participants presented their latest academic research to the group, covering visual and computational methods to address a wide range of geographic dynamics and application domains. The government panel and academic research briefings set the stage for animated breakout-group discussions. In the final session of the workshop, suggestions were made on how to sustain the momentum of these discussions as well as advance GIScience research on geographic dynamics. These interactions between UCGIS and government participants are helping to build a consensus view of the challenges and possibilities and to advance visualization and computation techniques and theory necessary for the understanding of dynamics in geographic domains.

ABOUT THIS BOOK

Authors who attended the workshop, as well as other authors who are working on topics relating to dynamics in geographic domains, were invited to submit proposed chapters for this book. Members of the workshop steering committee, as well as members of the broader geographic information science community, peer reviewed these chapters. Eleven chapters were finally selected for publication in this book.

The book is divided into three parts. This organization reflects the general topical grouping of presentations and discussions at the workshop. Part 1, "Cognitive Aspects, Representation, and Data Models," focuses on some of the conceptual and cognitive underpinnings of geographic dynamics in addition to some of the data-modeling topics that are related to understanding dynamics of geographic domains. The first chapter, by Colin Ware, is on reasoning about dynamic geospatial data from the perspective of human visual cognition. In this chapter, Ware considers some of the visual and cognitive challenges relating to the task of finding novel spatio-temporal patterns that occur infrequently in dynamic geographic domains and suggests how the creation of spatial patterns can be used to support this kind of task. Michael Goodchild and Alan Glennon consider the perspective of GIScience with respect to geographic dynamics and, in particular, the role of fields and objects as a framework for modeling dynamics. They identify five areas, or gaps, where current knowledge and technique falls far short of what is needed if GIScience is to provide general support for geographic dynamics. In the last chapter of Part 1, Steven Prager discusses how network representations, complex network theory, and related concepts can be used to advance comprehension of dynamic geographic domains. This chapter explores several ideas relating to the use of networks to understand dynamic geographies — for example, the role of networks and dynamic transactions or flows in a domain.

Part 2, "Analyses, Computation, and Modeling" presents research where the analysis of geographic dynamics is the focus and computational and modeling approaches are discussed. In the first chapter of Part 2, Daniel Goldberg, John Wilson, and Craig Knoblock consider methods for designing next-generation gazetteers and geo-coders for dynamic domains where geographic entities are continually changing and spatiotemporal contexts (not just the more typical spatial context) become the focus. This chapter shows how these tools together facilitate the integration of disparate, dynamic data sources for use in the analysis and interpretation of the changing world. Jaymie Meliker discusses the current state of the art regarding epidemiological methods for evaluating geographic exposures and hazards with a particular focus on the role of dynamics. In this chapter, Meliker describes exposure-reconstruction procedures that are not possible through "space only" GIS. Mei-Po Kwan and Fang Ren discuss modeling human activities and movements in space and time in an urban context. They present a time geographic perspective as a framework for this work, and discuss some of the 3D geovisualization and geocomputational methods that they have been developing as part of their research on human activity patterns. The final chapter of this part by Nina Lam, Guiyun Zhou, and Wenxue Ju introduces the use of spatial metrics as an approach to detecting changes from time-series images in order to help reduce extraneous effects and detect the real changes captured by imagery. Two examples linking visual changes with spatial metrics are presented based on images from a digital camera and Landsat-TM images of New Orleans, Louisiana.

Part 3, "Visualization and Simulation," focuses on recent research efforts on topics of visualization and simulation of geographic dynamics. Narushige Shiode and Li Yin describe their work to design and develop dynamic, 3D city models that capture the dynamics of the urban growth. They present a prototype model based on Buffalo, New York, to show how temporal transitions in urban areas can be accommodated with these models. The second chapter, by Thomas Butkiewicz, Remco Chang, William Ribarsky, and Zachary Wartell, discusses visualization approaches that capture the dynamics associated with urban environments and account for changes in these areas. Two types of applications are presented: terrain analyses, and urban planning that involves both long-term planning and large-scale urban projects. The developed approaches are tested on 3D urban models with possibly hundreds of thousands of structures leading to insights into urban morphology and how to effectively organize the dynamics of urban environments. David Bennett and Wenwu Tang present research on the construction of agent-based models that simulate the behavior of mobile, aware, and intelligent individuals, which interact with each other and the environment in which they live. A case study based on the migratory behavior of elk in Yellowstone National Park's northern range illustrates the challenges of simulating individualized and localized movements in a dynamic geographic domain. In the last chapter of this part, Narushige Shiode and Paul Torrens examine the manner in which urban space is developed and populated in a virtual city, AlphaWorld, in order to determine the extent to which this virtual city resembles its real-world counterpart, in this case, Austin, Texas. The comparison of virtual and real-world urbanization leads to a better understanding of the prospects for measuring the patterns of urban development in both types of environments and possible insights into future trends in urbanization.

REFERENCES

Batty, M. 2005. *Cities and Complexity.* Cambridge, MA: Massachusetts Institute of Technology.

Bennett, D. A. and W. Tang. 2006. Modelling adaptive, spatially aware, and mobile agents: Elk migration in Yellowstone. *International Journal of Geographical Information Science, Spatial Agent-based Modelling* 20(9):1039–66.

Campos, J. and K. Hornsby. 2004. Temporal constraints between cyclic geographic events. *Proceedings of GeoInfo 2004*, Campos do Jordao, Brazil, November 22–24, 2004.

Chen, J. and J. Jiang. 2000. An event-based approach to spatio-temporal data modeling in land subdivision systems. *GeoInformatica* 4:387–402.

Cole, S. and K. Hornsby. 2005. Modeling Noteworthy Events in a Geospatial Domain. *Proceedings of the First International Conference on Geospatial Semantics*, GeoS 2005, 78–92. Lecture Notes in Computer Science, 3799. Berlin: Springer.

Egenhofer, M. and D. Wilmsen. 2006. Changes in topological relations when splitting and merging regions. In *Progress in Spatial Data Handling—12th International Symposium on Spatial Data Handling*, ed. A. Riedl, W. Kainz, and G. Elmes, 339–52. Berlin: Springer.

Frank, A. 1998. Different types of "times" in GIS. In *Spatial and Temporal Reasoning in Geographic Information Systems,* ed. M. Egenhofer and R. Golledge, 40–62. New York: Oxford University Press.

Galton, A. 2003. Desiderata for a spatio-temporal geo-ontology. In *Proceedings of the International Conference on Spatial Information Theory,* COSIT 2003, ed. W. Kuhn, M. F. Worboys, and S. Timpf, 1–12. Lecture Notes in Computer Science. Kartause Ittingen, Berlin: Springer.

Galton, A. 2004. Fields and objects in space, time, and space-time. *Spatial Cognition and Computation* 4:39–68.

Galton, A. 2005. Dynamic collectives and their collective dynamics. In *Spatial Information Theory: Proceedings of International Conference on Spatial Information Theory,* COSIT 2005, ed. A. G. Cohn and D. M. Mark, 300–15. Berlin: Springer.

Galton, A. and M. Worboys. 2005. Processes and events in dynamic geo-networks. In *Geo-Spatial Semantics: Proceedings of First International Conference*, GeoS 2005, ed. M. A. Rodríguez, I. F. Cruz, S. Levashkin, and M. J. Egenhofer, 45–59. Lecture Notes in Computer Science, vol. 3799. Berlin: Springer.

Goodchild, M. F., M. Yuan and T. J. Cova. 2007. Towards a general theory of geographic representation in GIS. *International Journal of Geographic Information Science* 21(3):239–60.

Grenon, P. and B. Smith. 2004. SNAP and SPAN: Towards dynamic spatial ontology. *Journal of Spatial Cognition and Computation* 4(1):69–103.

Hornsby, K. and M. Egenhofer. 1997. Qualitative representation of change. In *Proceedings of the International Conference on Spatial Information Theory*, ed. S. Hirtle and A. Frank, 15–33. COSIT '97, Lecture Notes in Computer Science, vol. 1329. Berlin: Springer Verlag.

Hornsby, K. and M. Egenhofer. 2000. Identity-based change: A foundation for spatio-temporal knowledge representation. *International Journal of Geographical Information Science* 14(3):207–04.

Hornsby, K., M. Egenhofer, and P. Hayes. 1999. Modeling cyclic change. In *Advances in Conceptual Modeling, Proceedings of the ER'99 Workshops*, ed. P. Chen, D. Embley, J. Kouloumdjian, S. Liddle, and J. Roddick, 98–109. Paris, France, vol. 1727, Lecture Notes in Computer Science. Berlin: Springer-Verlag.

Huang, B. and C. Claramunt. 2005. Spatiotemporal data model and query language for tracking land use change. *Transportation Research Record* 1902:107–13.

Kwan, M.-P. 2001. Interactive geovisualization of activity-travel patterns using three dimensional geographical information systems: A methodological exploration with a large data set. *Sage Urban Studies Abstracts* 29(1):3–135.

Kwan, M.-P. 2004. GIS methods in time-geographic research: Geocomputation and geovisualization of human activity patterns. *Geografiska Annaler, Series B: Human Geography* 86(4):267–80.

Langran, G. 1992. *Time in Geographic Information Systems*. Bristol, PA: Taylor & Francis Inc.

Laube, P., S. Imfeld, and R. Weibel. 2005. Discovering relative motion patterns in groups of moving point objects. *International Journal of Geographic Information Science* 19(6):639–68.

Marceau, D. J., L. Guindon, M. Bruel, and C. Marois. 2001. Building temporal topology in a GIS database to study the land-use changes in a rural-urban environment. *The Professional Geographer* 53:546–58.

McIntosh, J. and M. Yuan. 2005. Assessing similarity of geographic processes and events. *Transactions in GIS* 9(2):223–45.

Miller, H. J. 2005. Necessary space-time conditions for human interaction. *Environment and Planning B: Planning and Design*, 32:381–401.

Parker, D., S. M. Manson, M. A. Janssen, M. J. Hoffmann, and P. Deadman. 2003. Multiagent systems for the simulation of land-use and land-cover change: A review. *Annals of the Association of American Geographers* 93(2):314–37.

Peuquet, D. 1994. It's about time: A conceptual framework for the representation of temporal dynamics in geographic information systems, *Annals of the Association of American Geographers* 84(3):441–61.

Peuquet, D. 2002. *Representations of space and time*. New York: Guilford.

Peuquet, D. J. and N. Duan. 1995. An event-based spatiotemporal data model (ESTDM) for temporal analysis of geographical data. *International Journal of Geographical Information Systems* 9(1):7–24.

Sriti, M., R. Thibaud, and C. Claramunt. 2005. A fuzzy identity-based temporal GIS for the analysis of geomorphometry changes. *Journal of Data Semantics* (Springer Verlag) 3(1):81–99.

Stewart Hornsby, K. and S. Cole. 2007. Modeling moving geospatial objects from an event-based perspective. *Transactions in GIS* 11(4):225–43.

Worboys, M. 1994. A unified model of spatial and temporal information. *Computer Journal* 37(1):26–34.

Worboys, M. 2001. Modelling changes and events in dynamic spatial systems with reference to socio-economic units. In *Life and Motion of Socio-Economic Units*, ed. A. Frank, U. Raper, J. Cheylan, 129–38. ESF GISDATA series, no. 8, Taylor and Francis.

Worboys, M. 2005. Event-oriented approaches to geographic phenomena. *International Journal of Geographical Information Science* 19(1):1–28.

Worboys, M. and Hornsby, K. 2004. From objects to events. GEM, the geospatial event model. In *Proceeding of GIScience 2004*, ed. Egenhofer, M., C. Freksa, and H. Miller, 327–43. Lecture Notes in Computer Science, 3234. Berlin: Springer.

Yuan, M. 1996. Modeling semantical, temporal, and spatial information in geographic information systems. *Geographic Information Research: Bridging the Atlantic*. ed. M. Craglia and H. Couclelis. London: Taylor & Francis: 334–47.

Yuan, M. 1999. Use of a three-domain representation to enhance GIS support for complex spatiotemporal queries. *Transactions in GIS* 3(2):137–59.

Yuan, M. 2001. Representing complex geographic phenomena in GIS. *Cartography and Geographic Information Science* 28(2):83–96.

Yuan, M. and K. Stewart Hornsby. 2007. In press. *Computation and visualization for understanding dynamics in geographic domains*. Boca Raton: CRC Press.

Acknowledgments

Many people contributed to this book. Thanks to all the authors who wrote chapters. The steering committee for the workshop that assisted with reviewing selections, and they include Nina Lam, Harvey Miller, Jaiwei Han, and Alan MacEachren. Additional reviewers drawn from the geographic information science community are thanked for their help in reviewing submissions to this book. They include Sean Ahearn, Michael Batty, Helen Couclelis, Thomas Cova, Mark Gahegan, Linda Hill, Werner Kuhn, David Mark, Harvey Miller, and Lynn Usery. Jack Sanders (Executive Director of the UCGIS), John Wilson (past President of the UCGIS), and Sean Ahearn, (current UCGIS President) all assisted with different aspects of the workshop organization and getting this book off the ground. Special thanks are given to Arnold Landvoight, who worked on many of the Intelligence Community-related aspects of the workshop and really helped with the coordination between the different communities. Assistance from the publisher is also gratefully acknowledged. Kathleen Stewart Hornsby's research is supported in part by a grant from the U.S. National Geospatial-Intelligence Agency HM1582-05-1-2039. May Yuan's effort is in part supported by the U.S. National Science Foundation through Collaborative Awards BCS-0416208.

Contributors

David A. Bennett
Department of Geography
University of Iowa
Iowa City, Iowa

Thomas Butkiewicz
Charlotte Visualization Center
University of North Carolina
Charlotte, North Carolina

Remco Chang
Charlotte Visualization Center
University of North Carolina
Charlotte, North Carolina

Alan Glennon
Department of Geography
University of California
Santa Barbara, California

Daniel W. Goldberg
Department of Geography
University of Southern California
Los Angeles, California

Michael F. Goodchild
Department of Geography
University of California
Santa Barbara, California

Wenxue Ju
Department of Geography and
  Anthropology
Louisiana State University
Baton Rouge, Louisiana

Craig A. Knoblock
Information Sciences Institute
University of Southern California
Los Angeles, California

Mei-Po Kwan
Department of Geography
The Ohio State University
Columbus, Ohio

Nina Lam
Department of Environmental Studies
Louisiana State University
Baton Rouge, Louisiana

Jaymie R. Meliker
Graduate program in public health
Stony Brook University
Stony Brook, New York

Steven D. Prager
Department of Geography
University of Wyoming
Laramie, Wyoming

Fang Ren
MS GIS Program
University of Redlands
Redlands, California

William Ribarsky
Charlotte Visualization Center
University of North Carolina
Charlotte, North Carolina

Narushige Shiode
Department of Geography
University at Buffalo (SUNY)
Buffalo, New York

Wenwu Tang
Department of Geography
The University of Iowa
Iowa City, Iowa

Paul M. Torrens
School of Geographical Sciences
Arizona State University
Tempe, Arizona

John P. Wilson
Department of Geography
University of Southern California
Los Angeles, California

Colin Ware
Data Visualization Research Lab
Center for Coastal and Ocean Mapping
University of New Hampshire
Durham, New Hampshire

Li Yin
Department of Urban and Regional
   Planning
University at Buffalo (SUNY)
Buffalo, New York

Zachary Wartell
Charlotte Visualization Center
University of North Carolina
Charlotte, North Carolina

Guiyun Zhou
Gannett Fleming, Inc/GeoDecisions
St. Louis, Missouri

Part I

Cognitive Aspects, Representation, and Data Models

1 Why Do We Keep Turning Time into Space?

Colin Ware

CONTENTS

Most visualization of geospatial data is done via the production of static images, although occasionally *animations* are produced to illustrate a temporal sequence, such as the growth in population, or some change in vegetation over time. The advent of commodity high-performance computer graphics makes it possible to produce animations at a low cost and high quality for almost any time-varying data set. But should we do it?

In this brief paper, I consider the problem of reasoning about dynamic geospatial data from the perspective of human visual cognition and argue that turning time into space is almost always the right thing to do. I illustrate my argument with some recent work analyzing the underwater behavior of humpback whales.

The most important reason for visually displaying data, as opposed to using words or tables, is to support *visual thinking*. Visual thinking is a distributed cognitive process where visual props, such as objects in the world, or graphical computer displays, support problem solving. As a number of theorists have noted, the key advantages of visual thinking are twofold: *memory support* and *problem solving through pattern finding* (Ware 2004).

Memory Support. Graphical entities such as images, symbols, and patterns on a display provide proxies for concepts. When these entities are fixated, the corresponding concepts become activated in the brain. This kind of visually triggered activation can often be much faster than the retrieval of that same concept from long-term memory in the absence of graphical aids (for example, with the eyes shut). When an external-concept proxy is available in the form of a visual symbol, access is made by means of eye movements that typically take approximately one-tenth of a second. Once the proxy is fixated, a corresponding concept is activated within less than two-tenths of a second, hence the advantage over long-term memory retrieval, which can take several seconds, or more. It is possible to place upward of 30 concept proxies (in the form of images, symbols, or patterns) on a screen, providing a very rapidly accessible concept buffer. Compare this to the approximately three concept "chunks" that we can hold in verbal working memory at a time.

But there is a major limitation. For visual proxies to trigger concept activation, previously learned associations must exist between the visual symbols, images, or patterns and certain concepts. Now consider the problem of an *animated* replay of time-varying geospatial data. Concept proxies will disappear and reappear only when the relevant time window is current on the display. To get the advantage of memory support, it is necessary to either watch a replay or use some fast-forwarding device to get to a point where the symbol appears.

Problem solving through pattern finding. When we are using a visual display for analysis, the goal is generally to find patterns in the data. These may be patterns having an already known characteristic form; for example, the spatial signature of deep rock formations signals the likely presence of trapped gas or oil to a geologist. Or they may be previously unknown patterns. Finding a novel pattern can be an important act of discovery.

We can perceive temporal patterns as well as spatial patterns. For example, non-verbal communication through hand gestures has much more to do with the temporal sequence of hand positions than with the actual shape of the hand. The problem is that for temporal patterns to be perceived clearly they must occur in a short interval. We perceive the movements of the second hand on a clock face, but not the minute or hour hands. Studies of human *visual working memory* suggest that for a motion pattern to be readily perceived, it should occur within an interval of two seconds or less. Temporal patterns will be clearest if they can be perceived within a single fixation of perhaps half a second.

This, obviously, poses a problem for those seeking to find novel spatiotemporal patterns that occur *infrequently* in time-varying geospatial data. The analyst watching an animated replay of such data must resort to formally encoding candidate patterns in long-term memory. This usually means constructing a verbal description of the pattern and remembering it or writing it down. When a pattern reoccurs, it can then be compared to the stored pattern. Not only is this cognitively effortful but our capacity to encode novel patterns is very limited. We may only be able to encode half a dozen temporal patterns in an hour of video, and these may not be the right ones.

As an alternative, if the spatiotemporal pattern can somehow be turned into a spatial pattern, then visual comparisons between sections of a time sequence can be made by means of eye movements. Because eye movements are so fast, this enables several patterns a second to be compared.

Resolution is concentrated at the fovea as illustrated in Figure 1.1. Our ability to see detail is really captured only in the central region of vision, and so in order to find patterns we make eye movements from point to point. Eye movements typically occur at a rate of one to four per second.

It is useful to classify eye movements as a kind of navigation in space to access new information, and it is useful and instructive to compare eye movements with

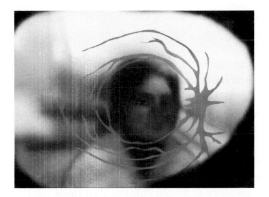

FIGURE 1.1   This illustration shows the quality of the visual information as it varies across the visual field. The spidery object is the pattern of blood vessels that actually lie on top of the receptors. We are able to see far more detail with a central region of the retina, called the fovea, which is only the size of a thumbnail held at arm's length. Thus the process of seeing involves sampling the world by moving the eye rapidly from point to point, gathering critical task-relevant information.

other forms of navigation for information access. If we are working on a paper, walking to the library to get additional information should be considered a cognitive cost. This cost can be compared to conducting an online search via a Web browser that will generally consume less time and cognitive effort. A selection of navigation methods is offered in Table 1.1 for comparison.

TABLE 1.1
Approximate Times for a Selection of Perceptual/Cognitive Acts

Eye movements	0.1 seconds	Tightly coupled with visual cognitive system.
Walk	4 kph/dist (minutes)	Good affordances for contemplative thought. Poor affordances for interactive cognitive work.
Drive	80 kph/dist (minutes-hours)	Poor affordances for interactive cognitive work. Moderate affordances for contemplative thought.
Zooming	Dist = 4 t (seconds)	Only used interactively. The most efficient way of traversing large distances with visual continuity.
Mouse select	1.5 sec	A mouse selection of a Web link can cause an entirely new page of information to appear.
Playback for time-varying material, e.g., movies	Time is = length of the material	Under some circumstances playback can be speeded up by a factor of three or four.

Interactive computer interfaces allow navigation, through space and time to different parts of the data space. Depending on the design of the user interface a typical act of spatial navigation using a computer interface takes a few seconds. Although this is not as fast as navigation by making an eye movement, computer-based navigation is generally much faster than most other methods for seeking information.

Patterns are held in various memory buffers in the brain. Four of these are relevant here. *Visual working memory* holds at most three very simple patterns for a period of usually only one or two fixations. However, with greater cognitive effort informative patterns can be held in visual working memory for a few seconds. This allows us to make simple visual pattern comparisons. Patterns are held in visual working memory while we execute eye movements.

Visual long-term memory can hold information indefinitely. However, the rate at which we can store new information is very low; even simple new patterns can be acquired at a rate of only roughly one every few minutes.

Verbal working memory holds verbally coded information for a few seconds. If a visual pattern can be given a verbal label it can be used to extend visual working memory.

Verbal long-term memory can hold information indefinitely. Moreover, as with visual long-term memory, only a small amount of information can be retained.

Note taking provides a means of retaining information indefinitely; however, it disrupts the cognitive workflow, and retrieval can be difficult. Note taking can be done with either sketches or words.

Although the cognitive model we have described so far is an extreme simplification of the actual processes that occur in human cognition, it is sufficient to help us understand fundamental differences between different types of user interface used for analyzing spatiotemporal data.

A SIMPLE COGNITIVE MODEL OF PATTERN DISCOVERY

To illustrate the value of approximate, first-order cognitive modeling in user interface design, the remainder of this chapter is devoted to an extended example involving the analysis of the underwater behavior of humpback whales to find novel behavior patterns. We were especially interested in finding stereotyped behaviors of whales. Thus we sought behavior patterns that were repeated many times.

We show how two user interfaces differ dramatically in terms of their cognitive efficiency and how this difference can be predicted using the very simple cognitive two-step process model that follows.

> *Step 1. Conduct a visual search for an interesting (e.g., unusual) visual pattern. Once an example is found, store it in memory.*
> *Step 2. Conduct a visual search for similar visual patterns in the display.*
> *Repeat steps 1 and 2 until no new patterns are found.*

The key to executing this efficiently is that visual search can be executed rapidly, and reliance on memory should be minimized.

FIGURE 1.2   The tag, shown below, is attached to the back of a humpback whale by means of suction cups. The tag remains attached for up to 24 hours, recording the orientation and depth of the whale continuously. Once the tag detaches it is retrieved and the data is downloaded. This data is used to construct a pseudotrack showing the 3D trajectory of the whale together with the whale's orientation in space.

FIGURE 1.3   DTAG records accelerometer and magnetometer data as well as depths and sound (Johnson and Tyack 2003).

EXAMPLE: FINDING NOVEL BEHAVIOR PATTERNS FROM WHALE-TRACK DATA

We (Ware et al. 2006) were faced with the problem of visualizing data coming from a tag, attached to a whale with suction cups, that could provide several hours of data on the position and orientation of the whale as it foraged for food at various depths in the ocean (see Figures 1.2 and 1.3).

INTERFACE 1: MOVEMENT PLAYBACK WITH NOTE TAKING

Our first attempt to provide visualization support involved a tool that allowed ethologists to replay the motion of the whale at any desired rate. This was based on our GeoZui4D system, and its user interface is illustrated in Figure 1.4. Hours of study using this tool resulted in the identification of some interesting behavioral patterns. But, although there was some obvious utility, analysis was very time-consuming.

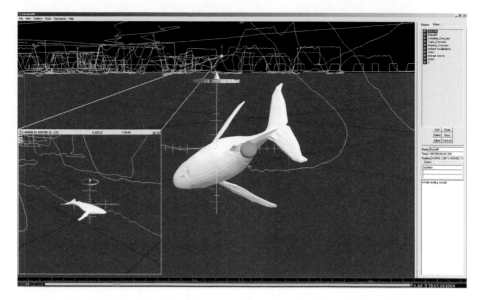

FIGURE 1.4   The GeoZui4D interface is shown being used to analyze whale behavior. The buttons on the right-hand side are space-time notes. They capture a view, a time, and an annotation. **(See color insert after p. 110.)**

COGNITIVE ANALYSIS FOR METHOD 1

In method 1, an ethological analysis involved an animated replay over time.

The ethologist observed the movements of the whale and attempted to identify whether a particular form of motion was stereotyped. This was a multipass process in which each pass took almost as much time as the time that elapsed during tagging. Using the playback method it was possible to identify only two or three stereotyped patterns in the course of a playback session that might take several hours. In order to see if these same behaviors occurred with other whales it was necessary to play back the data from each of the other tagged whales. We can now fill in some of the details in the model.

> **Step 1.** *Conduct a visual search for interesting (e.g., unusual) visual patterns by continuously observing while **playing back the behavior.** Once an example is found store it in **long-term memory.***
> **Step 2.** Conduct a visual search, **using playback,** for similar visual patterns in the display.
> **Repeat steps 1 and 2 until** no new patterns are found.

If at any time a new behavior was discovered in any whale's data, it was necessary to revisit all of the whale-track data to see if that new behavior occurred elsewhere. Therefore, all the data would have to be viewed several times in order to do a thorough analysis. If we had 80 hours of tag data it might take 80×5, or 400 hours to do a thorough analysis by this method.

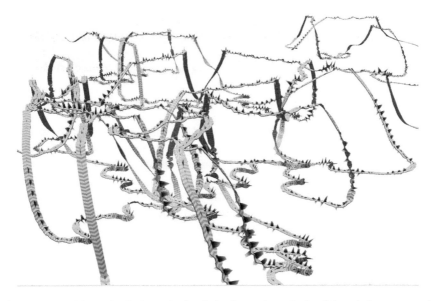

FIGURE 1.5   This track of a humpback whale shows the attitude of the whale over a period of a few hours. The sawtooths indicate where fluking occurred. The yellow portions of the track show where the roll angle exceeded 40 degrees from horizontal. It is immediately evident that a highly stereotyped behavior is occurring at the bottom of the dives where the animal repeatedly rolls on its side for a fixed duration before righting itself. **(See color insert after p. 110.)**

In fact, we developed a number of tools to speed up the analytic process. Chief among these was space-time notes, which allowed us to mark interesting behaviors and rapidly return to them later. Although these helped improve the quality of the analysis that could be done, they probably only reduced the analysis time by a factor of two at most.

METHOD 2: RIBBON PLOT WITH RAPID SPATIAL NAVIGATION

Our second attempt involved creating a pseudo-track of the whale that represented the attitude of the whale, as well as its path using a flattened ribbon (see Figure 1.5). This was embedded in a second application that we called TrackPlot. TrackPlot allowed rapid track-centered navigation through the data set. It also allowed for zooming in and out to reveal patterns at different scales, but the key innovation was the ribbon representation.

With the aid of TrackPlot, it often took less than a minute with each new set of whale data to find new patterns in the data, and to recognize where the whale was exhibiting behaviors we had already observed with other animals. We used about thirty seconds of this time fast-forwarding at high speed to an interesting patch of behavior. Following this, we carried out the pattern-finding activity by means of a visual search using *eye movements*. Our cognitive model now looks like this.

Step 1. *Conduct a visual search for an interesting (e.g. unusual) visual pattern* **using eye movements.** *Once an example is found store it in* **visual working memory.**

Step 2. *Conduct a visual search,* **using eye movements,** *for similar visual patterns in the display.*

Repeat steps 1 and 2 until *no new patterns are found.*

Note that with this second solution working memory has been substituted for long-term memory, and eye movements have been substituted for playback as a method of pattern seeking. Each of the search steps might require ten fixations and take a few seconds to cover the entire display, since eye movements can be carried out several times a second. This means that the entire two-step process could be repeated as many as a dozen times within a minute. Using the TrackPlot ribbon display it took minutes rather than hours to carry out a preliminary analysis of the data.

This example shows that transforming a temporal pattern into a spatial pattern can result in an enormous gain in cognitive efficiency. The rate at which new whale behaviors could be discovered was increased by at least two orders of magnitude. More generally, our analysis suggests that except in the case of very rapid, temporal patterns, the best way of presenting spatiotemporal data will not be in the form of a playback of events. Instead, the data should, wherever possible, be transformed into spatial patterns in a way that supports rapid visual search.

CONCLUSION

Whale behavior analysis is a very specialized activity. But in many cases the analysis of time varying geospatial data requires a search for recurring patterns. These will suffer from the same problem as whale behavior analysis if playback is used for the search.

In the case of the whales, the key to the new method of analysis was to find a way of transforming motion patterns into spatial patterns, with a method of representation that could be intuitively "read." The purpose of the simple cognitive analysis given here has been to suggest that turning time varying events into spatial patterns should be a general strategy when we are confronted with the need to analyze time varying data. Only as a last resort should data be replayed for analysis, because of the enormous investment in time that is involved.

But there are some obvious cases when this rule of thumb does not apply. The goal in the whale study was exploratory data analysis to find novel stereotyped patterns. Visualizations may also be used as a way to help describe patterns that have already been discovered. The use of visualization in explanation is very different from its use in analysis, and the cognitive model does not apply. The goal of an explanatory videotape is not visual efficiency, but the creation of a visual narrative that carries the viewer as a passive participant. The only cognitive activity involved is the acquisition of predigested information, rather than the creation of new knowledge. For this reason, animated replay can often be very effective in pedagogical applications when the parts of the data that are shown are always carefully selected for their informative value.

Of course, turning time into space is not a novel idea. Even a simple trend plot, showing stock market values over time, does this. Our purpose in this paper has been to develop a cognitive argument for why this is the right thing to do.

REFERENCES

Johnson, M. P. and P. Tyack. 2003. A digital recording tag for measuring the response of wild marine mammals. *IEEE J. Oceanic Eng.* 28(1):3–12.

Ware, C., R. Arsenault, M. Plumlee, and D. Wiley. 2006. Visualizing the underwater behavior of humpback whales. *IEEE Computer Graphics and Applications*, July/August: 14–18.

Ware, C. 2004. *Visual Thinking: Perception for Design*. San Francisco: Morgan Kaufman.

2 Representation and Computation of Geographic Dynamics

Michael F. Goodchild and Alan Glennon

CONTENTS

INTRODUCTION: THE DOMAIN OF GEOGRAPHIC DYNAMICS

In the parlance of computer-aided software engineering (CASE), one designs solutions to problems in a domain by identifying a series of representative *use cases*. Success depends on choosing a sufficient number of use cases to sample the domain adequately, characterizing the variety of likely applications of the system. By this logic, the design of systems to represent and compute geographic dynamics requires an understanding first of the domain of geographic dynamics, and second of the range of uses to which such representations and computations will be put. In short, to discuss representation and computation of geographic dynamics we must first understand the full range of geographic dynamics, and then ask why such representations and computations are useful.

The adjective *geographic* refers primarily to the surface and near-surface of the earth. For many purposes the two-dimensional surface is sufficient, but atmospheric scientists, geologists, and mining engineers are also interested in areas above and below the surface, and in full three-dimensional knowledge about these domains. Thus the geographic domain can be defined as the roughly 500 million sq km of the surface, together with the first 20 km or so above the surface, and the first 20 km or so below it. Spatial resolution within this domain is a little harder to quantify, but there is little significant interest in resolutions coarser than 10 km, or in resolutions finer than 10 cm.

Dynamics refers to change through time, and the characterization, understanding, and prediction of such change. Change in the geographic world can result from naturally occurring processes, such as erosion, or human-induced change, such as global warming. Dynamic phenomena extend from the daily journeys to work made by commuters to the changes of land cover induced by wildfire or severe storms, to tides and currents, to changes in land use as a result of urban development. In the traditions of cartography and geographic information science (GIScience), such phenomena have been difficult and expensive to record and store on maps or in databases, and geographic information systems (GIS) have often been criticized for not accommodating knowledge about the dynamic aspects of the earth's surface. Topographic mapping practice tends to emphasize the relatively static aspects of the surface, such as terrain, hydrography, and built form, and such practice has been inherited by GIS, which were developed originally in large part as systems for storing the contents of maps (Goodchild 1988).

Defined in this way, the domain of geographic dynamics is clearly vast, since it spans the concerns of a large number of disciplines that include geography but also virtually any discipline concerned with change in geographic space: geology, atmospheric science, ecology, economics, criminology, and many more. In recent years a wide range of tools have been developed to address this domain, including tools for visualization, simulation of the actions of agents, the operation of cellular automata, the solution of partial differential equations, and much more. PCRaster (http://pcraster.geo.uu.nl; Burrough, Karssenberg, and van Deursen 2005) is perhaps the most prominent example of a GIS designed specifically for dynamics, and though as its name suggests its functions are primarily in the raster domain, a stunning array of examples have been explored, and it is clear that remarkably convincing simulations of processes operating in the geographic domain can be simulated with very simple rules. Examples range from the growth of volcanoes and the erosion of fault-block topography to seed dispersal and the movement of groundwater.

Also in the raster domain, researchers have used simple rules to simulate the changes of state that occur in urban growth and other changes of land use. These models are best seen as examples of cellular automata, made popular by John Conway. Clarke's SLEUTH model is of this type (Clarke, Hoppen, and Gaydos 1997) and has been applied to the modeling of urban growth in several areas of the United States, and similar models have been described (see, for example, White, Straatman, and Engelen 2004). In atmospheric science modeling has reached a high level of sophistication. At the global scale, a number of global climate models (GCMs) have been built to provide numerical solutions to the partial differential equations governing the atmosphere, and similar models have been constructed at the mesoscale to address local atmospheric phenomena. The modeling of tidal movements and ocean currents has also reached a high level of sophistication.

Agent-based models attempt to simulate the movements and actions of individual, autonomous agents and have had success in the study of the behavior of pedestrians in cities (Batty 2005), tigers in India (Ahearn and Smith 2005), and vehicles on congested highways. The use of GPS (the Global Positioning System) to track samples of individuals in cities has led to useful new knowledge about travel behavior (Kwan and Lee 2004). Finally, much progress has been made in the modeling of severe storms and their impacts (Yuan 1999, 2001).

THE ROLE OF GISCIENCE

What role should GIScience play within this vast and complex domain? Modeling of dynamic geographic phenomena is well established in many disciplines ranging from atmospheric science to hydrology and transportation. Just as GIS attempts to provide a generic set of tools to support the analysis and manipulation of geographic information, we propose that GIScience should similarly address the generic spatial components of dynamics, by devising answers to such questions as the following: What are the common aspects of modeling that span the entire domain of geographic dynamics? What generic tools can be designed to support the domain? What languages might provide generic support, allowing models in a wide range of application areas to be defined in a common, interoperable syntax? Is it possible to conceive of a generic data model that specializes the concepts of GIS to the needs of geographic dynamics, and in turn can be specialized to the needs of specific application domains? What general properties are exhibited by dynamic geographic phenomena that might constitute laws of geographic dynamics comparable to Tobler's First Law (Tobler 1970; Sui 2004)?

This is not a simple charge, because it implies knowledge of the entire domain, and the ability to generalize from all of its aspects. If it has taken GIS 40 years to advance from its primitive beginnings, with prototypes that addressed very specialized applications, to the generic tools of today, then the task of addressing geographic dynamics is clearly at least as forbidding. But this pursuit appears to be the only rational way to approach the question of the appropriate role for GIScience.

The distinction that is often drawn between form and process (Goodchild 2004) seems to be a useful way to begin to structure this charge, and to create a conceptual framework within which it can be addressed. Form is defined as how the world *looks*, and clearly this definition resonates with the traditions of GIS, with the reliance on the map as the primary source of GIS data, and with the use of imagery representing instantaneous snapshots of the earth's surface. Spatial analysis has typically emphasized the power of such *cross-sectional* data to reveal useful insights into patterns of phenomena on the earth's surface. Tobler's First Law stands as a powerful generalization about geographic form, and provides the basis for spatial interpolation and the fields of spatial statistics and geostatistics.

Nevertheless, the concept of how the world looks can be readily generalized to the spatiotemporal case. Three-dimensional visualizations of tracks, for example, are snapshots that focus on form, albeit over finite ranges of both space and time, and provide potential insights into individual behavior. Thus a change of emphasis in GIScience from space to space-time does not necessarily imply a simultaneous change of emphasis from form to process, or from how the world looks to how the world *works*. The distinction between form and process is not so much a distinction between space and space-time as one between data and rules; between the data that describe the details of how the world looks, and the rules, equations, and algorithms that describe how it works, and how it is transformed from a state at time t_i to time t_{i+1}.

Such rules, equations, and algorithms are most useful if they apply everywhere in space and time. The Second Law of Thermodynamics or the Periodic Table of the Elements would be of little value if they applied only in Nebraska, for example, or only on Tuesdays. Such knowledge is termed *nomothetic*, to distinguish it from detailed knowledge about the unique properties of times and places, that is, *idiographic* knowledge. The scientific community is in no doubt about the comparative merit of nomothetic knowledge, and terms associated with idiographic knowledge, such as *descriptive* or *anecdotal*, can be distinctly pejorative. From a GIS perspective, the distinction between nomothetic and idiographic aligns closely with the distinction between the software — the methods, scripts, procedures, and algorithms — and the database of local detail to which the software is applied, and which it transforms.

In summary, then, the role of GIScience in this context of geographic dynamics is to find general structures that support the domain. These may take the form of algorithms, simulation models, data models, languages, standards, knowledge about the modeling and propagation of uncertainty, and so on. By looking for generic solutions, GIScience pursues structures that are sharable, formal and unambiguous, reusable, and thus efficient.

FIELDS AND OBJECTS

The distinction between continuous-field and discrete-object conceptualizations appears to lie at the most fundamental level of GIScience. Briefly, continuous fields map every location in space-time to a variable, $z = f(\mathbf{x})$ where z denotes a property and may be nominal-, ordinal-, interval-, or ratio-scaled, and may denote a scalar or a vector. Examples include elevation and soil class as scalar functions of the two horizontal dimensions, and wind speed and direction as a vector function of the three spatial variables and time. Discrete objects, on the other hand, represent a conceptualization of the geographic world as an empty space littered with points, lines, areas, or volumes, each having a set of homogeneous properties. Discrete objects may be persistent through time, and may change shape and move.

The field/object distinction provides a convenient framework for the discussion of geographic dynamics. Some processes are conceptualized entirely within the field domain. They include those described by partial differential equations (PDEs): the behavior of viscous fluids (the Navier-Stokes equation) and of groundwater (the Darcy flow equation), and electromagnetism (the Maxwell equations). PDEs are normally solved in space-time using one of two methods: finite differences (FD) and finite elements (FE). FD methods approximate the derivatives using differences in a simple raster, and are therefore readily supported by raster functions in GIS. Derivatives in time are approximated by taking differences between consecutive rasters. The routine GIS function of slope calculation from a DEM is a simple example of numerical approximation, giving an estimate of the derivative of the field with respect to the horizontal dimensions. FE methods use an irregular mesh, and while this bears some resemblance to the TIN (triangulated irregular network) of GIS, FE meshes commonly utilize both triangles and quadrilaterals, represent variation within elements using curvilinear functions, and require continuity of value, gradient, and curvature across element edges (Carey 1995). By contrast, TINs conventionally require only continuity of value and assume linear variation within elements. Thus the TIN model would be problematic for the solution of PDEs because gradients are undefined across edges. Moreover, some global climate models are operationalized entirely in the spectral domain, requiring none of the spatial discretizations common in GIS.

Because of their similarity to raster methods, FD solutions of PDEs can be implemented readily in GIS, particularly in PCRaster, which uses a command language (van Deursen 1995) that can easily accommodate such applications. Software for the simulation of cellular automata can also be adapted fairly readily to FD solutions. Packages such as PCRaster and ESRI's ModelBuilder (www.esri.com/software/arcgis/about/modelbuilder.html) will support the loops needed in any iterative algorithm. However, these packages adopt a very simple approach to data management, in which each time interval is maintained as a separate raster layer, and each time step requires the input and output of at least two complete layers. Much more sophisticated approaches to the storage of ordered stacks of rasters have been devised, notably in the interests of compressing video (www.mpeg.org), and could be implemented or adapted to improve the performance of iterative GIS algorithms.

The integration of FE methods with GIS has proceeded much more slowly, however, and to date FE meshes are not one of the representations of fields that are supported by the most popular products. Some efforts to integrate FE software with GIS have been reported, where the GIS is used primarily to prepare data, visualize results, and analyze results in geographic context. Lack of support for FE methods is one of the more obvious gaps in current GIS support of geographic dynamics.

While some processes are conceptualized entirely within the field domain, others are conceptualized as interactions between objects. Gravity provides one example, since the movements of celestial objects are modeled through the object-to-object forces defined by the inverse-square Law of Gravitational Attraction. Once the number of bodies exceeds two, the mathematics rapidly becomes intractable, and the Many-Body Problem is notorious for its complexity.

In the geographic domain, object-based concepts of process can be found in the Spatial Interaction Model (Fotheringham and O'Kelly 1989) and its applications to travel behavior, migration, and social interaction. Batty (2005) has modeled pedestrian flows during London's Notting Hill Carnival using simple rules of interobject interaction, and similar methods have been used to model animal behavior. Software support is provided by several comprehensive packages for modeling autonomous agents, including SWARM (www.swarm.org) and REPAST (http://repast. sourceforge.net), and the latter has been integrated with ESRI's ArcGIS in the Agent Analyst extension (www.institute.redlands.edu/agentanalyst/AgentAnalyst.html). ESRI's Tracking Analyst provides some basic capabilities for handling the space-time trajectories of objects, and recently Google Earth (http://earth.google.com) has announced an extension to handle moving objects, though its capabilities are focused on simple visualization.

In a recent paper Goodchild, Yuan, and Cova (2007) argue that the dynamic behavior of objects can be captured in three fundamental dimensions. The vertical dimension of Figure 2.1 represents object shape, and distinguishes between objects that retain shape through time and objects that change shape. For example, a vehicle generally behaves as a rigid body through time, whereas clouds tend to change shape rapidly. The dimension toward the viewer represents an object's internal structure, and distinguishes between objects that are homogeneous (as tradition in GIS demands) and objects that have internal variation that may also be changing in time. For example, severe storms have complex and evolving internal structures. Finally, the dimension away from the viewer represents movement, and distinguishes between objects that are fixed and those that move through time. The traditional GIS representation, with its static, homogeneous objects, is represented by the lower-left corner.

Implementation of object-based processes is also aided by developments in object-oriented data modeling, and specifically by the widespread adoption of Unified Modeling Language (www.uml.org) as a CASE tool for database design. In this environment it is easy to implement representations of dynamic phenomena, including events, transactions, and flows (see the next section), that would never have been possible under the earlier map-based conceptualization of geographic reality. Smooth

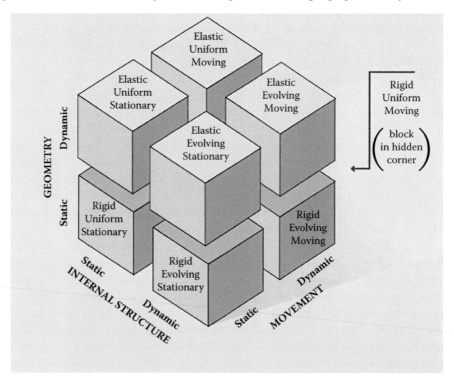

FIGURE 2.1 A representation of three fundamental dimensions of discrete-object dynamics (Goodchild, Yuan, and Cova 2007; reproduced by permission of Taylor and Francis).

integration of UML-based designs, developed in graphics environments such as Microsoft's Visio, with GIS packages such as ESRI's ArcGIS, has vastly improved our ability to represent dynamic phenomena in formal, readily sharable ways (Arctur and Zeiler 2004); this strategy is explored further in the next section, which focuses on the modeling of flows as an example. Other data-modeling developments include CityGML (www.citygml.org), an extension of GML (Geography Markup Language), which is itself an extension of XML (eXtensible Markup Language), that integrates the representations of 3D structures into standard GIS data models, using IFC, the international exchange standard of the construction industry.

Finally, some concepts of process combine both objects and fields. A ball rolling downhill provides a simple example, in which the ball (a discrete object) responds to the gradient of a field (elevation). Other examples include aircraft tracks through a field of wind, which are typically optimized to minimize travel time and fuel consumption, a home buyer looking for a suitable neighborhood and responding to continuous variation in perceived suitability, and a fugitive searching a landscape for concealment sites. Some relevant methods are included in current GIS, such as the calculation of geodesics (de Smith, Goodchild, and Longley 2007), but by and large the support for such processes is currently weak.

AN EXAMPLE: REPRESENTATION OF FLOWS

This section focuses on flows, in order to provide an example of how object-oriented data modeling can offer general, formal solutions to problems of representation in geographic dynamics. As noted at the outset, any representation that claims to be generic must demonstrate its applicability over all of a defined domain. More specifically, the data model must provide a *slot* for the storage of all information relevant to a particular application or use case; and use cases must be chosen to sample all of the defined domain.

We defined three use cases as representative of the domain of geographic flows. The first was a summary table of migrations between U.S. states in the period 1995 to 2000, as reported by the U.S. Bureau of the Census (Figure 2.2). The second was the famous Minard map of Napoleon's 1812 Russian campaign, showing the route of the army, major events of the campaign, and the steady diminution of the army's size from over 600,000 to little more than 10,000 (the map is celebrated by Tufte, 1983, as an example of very effective and economical visual display of information). The map is shown in Figure 2.3 as rerendered using modern GIS mapping tools (ESRI's ArcGIS). The third use case was the hydrology of part of the Central Kentucky Karst (Figure 2.4), an area of mixed surface and underground flow, where all of the surface routes and some of the underground routes are known and mapped, but where other underground routes have been inferred by dye tracing.

Taking these three use cases, we identified the classes of objects present in each case. For the migration data, these were the origin and destination objects, and the flows between them. We identified the relevant attributes of each class, and developed a UML diagram showing each class, relationships between the classes, the attributes of each class, and any methods associated with the class. For the Minard map, the various segments of the march formed one class, and the events along the

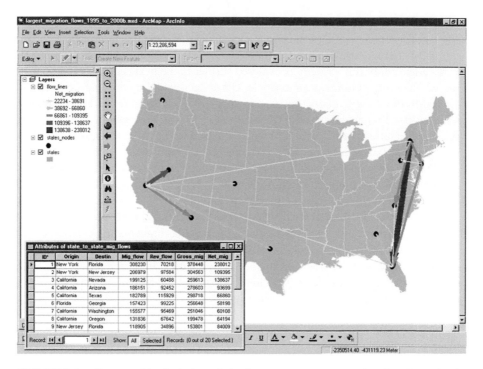

FIGURE 2.2 Cartographic depiction of the largest state-to-state migration flows in the period 1995–2000.

route another. The map shows the size of the army as a continuously changing attribute of the route, requiring a functional representation as a UML method, an option that is not widely implemented in GIS network software.

One of the aspects common to both the karst map and the migration data is the presence of inferred routes, where flow follows an unknown track. In the migration case, for example, we have no knowledge of the actual tracks followed by migrants, and in the karst case the unexplored underground routes of water are similarly unknown. Cartographically, such inferred paths are often shown as dashed, and depicted as simple straight lines (Figures 2.2 and 2.4).

Figure 2.5 shows the final merged result, in which each class is defined generically and given attributes that are common across all use cases. When the model is applied to any of the three use cases, each class is specialized to meet the context, and additional context-specific attributes are added. Starting from the left, each flow on the map forms a class, with an associated ID. Flows may be associated either with network reaches, which are real tracks on the earth's surface and associated with polylines, or with implied links. Implied links are associated with input and output nodes, which are associated with polylines through connections represented by the incidence class.

By giving flow phenomena this formal structure, we ensure that systems can be made interoperable, and that terms can be shared between widely divergent applications. The data model has been integrated with ESRI's ArcGIS, and a collection of

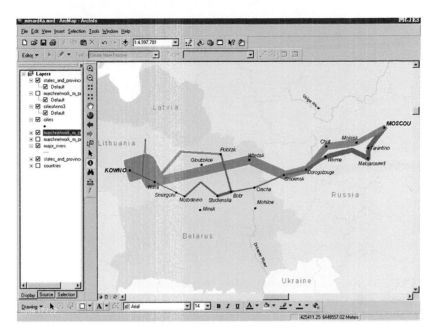

FIGURE 2.3   A re-rendering of the Minard map of Napoleon's 1812 Moscow campaign. **(See color insert after p. 110.)**

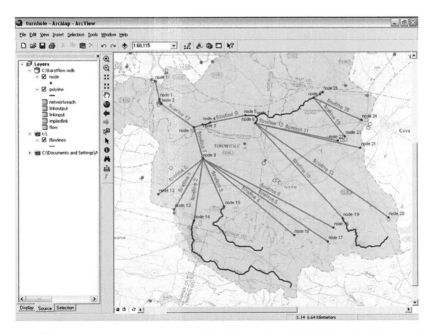

FIGURE 2.4   The hydrology of a part of the Central Kentucky Karst. Red lines indicate inferred flows. **(See color insert after p. 110.)**

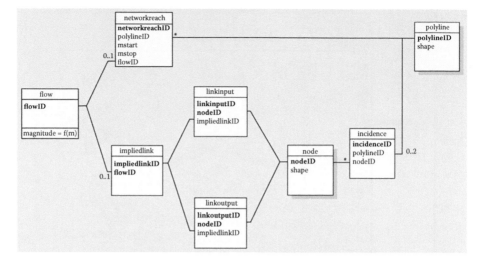

FIGURE 2.5   A generic data model for flow phenomena.

tools are available for help with populating the model, specializing it for applications, and building analysis scripts (http://dynamicgeography.ou.edu/flow/index.html).

Galton and Worboys (2005) also use flows in networks as an example of the representation of dynamic geographic phenomena, based on traffic in transportation networks as the use case. Their model also includes the notion of continuous change of attributes along links, which they term *seepage*. Their concept of dynamics extends to the construction of new links and the deletion of existing ones, but the need to deal with inferred links did not arise in their use case.

AN AGENDA FOR GISCIENCE

Although geographic dynamics is a vast domain, the simple conceptual framework presented above does at least allow the identification of significant gaps in our current abilities, and the basis for a research and development agenda. In this section we discuss five topics that in our view constitute significant gaps, and review the current state of the art with respect to each of them.

LANGUAGES FOR OBJECT DYNAMICS

From a GIS perspective, there is a conspicuous difference between the high level of development of languages for handling raster dynamics, and the comparable state of languages in the vector domain. As early as the late 1980s Dana Tomlin was developing Map Algebra, a simple synthesis of raster operations into four basic types (Tomlin 1990). PCRaster's scripting language, the subject of a monograph by van Deursen (1995), provides a much less verbose and more rigorously defined replacement, and a comprehensive underpinning for modeling field-based processes. It allows entire fields to be addressed through symbols, and defines a series of functional operators that cover most of the requirements of implementing cellular automata and FD approaches to PDEs. For example, the statement C = A + B directs the system to add each cell's value in raster A to the corresponding cell's value in raster B, to create a new discretized field C. As such, it provides a vastly simpler alternative to programming in the traditional source languages.

Despite this progress, and the popularity of implementations of Map Algebra in many GIS packages, the equivalents for vector representations of fields and for object dynamics have as yet failed to emerge. Kemp (1997a, b) and Vckovski (1998) have argued that the user interface to a GIS could be vastly simpler if fields could be addressed symbolically, independently of their discretization and spatial resolution. For example, the statement C = A + B might be executed even though A was represented as a raster and B as a TIN, the system making the necessary decisions about interpolation methods and the best representation for C (perhaps using A's raster and the average value of B within each cell, on the grounds that the spatial resolution of the raster is explicitly defined whereas the TIN's is not). In this approach the task of polygon overlay, long celebrated as the most daunting of GIS operations, would never be invoked explicitly, but triggered automatically whenever an operation required the mixing of two different spatial discretizations.

The basis for such a language might lie in relational algebra, given the power of object-oriented approaches in representing object dynamics. Clues to a solution might also be found in languages of object dynamics such as STELLA, which lack the focus on space-time but have powerful tools for representing interactions.

SOFTWARE OBJECTS FOR DYNAMICS

Over the past decade or so developments in software engineering have radically transformed the nature of GIS software, largely replacing monolithic packages with collections of reusable components. One software developer can now market many different products aimed at specific niches, while knowing that key code objects need to be developed only once. Functions implemented as components in GIS software can now be integrated with components from other packages, under the control of scripting languages such as VBA (Visual Basic for Applications; see, for example, Ungerer and Goodchild 2002) or Python (www.python.org).

Such strategies represent a high level of understanding of the domain, because they require software developers to identify the fundamental granules of data manipulation. In GIS there is significant consensus on this issue, but in the domain of geographic dynamics consensus appears to be largely absent. As we have argued, the domain of geographic dynamics is vast, spanning many disciplines. Bennett (1997) has reported significant progress on this issue, but more broadly, we still lack a clear consensus on the set of tasks that constitute computation of geographic dynamics, and the fundamental components into which those tasks can be decomposed. Instead, most efforts at modeling processes are implemented as stand-alone software in source languages such as C, with very little reuse of code. Similar comments can be made about the failure to date to achieve reusability in the coding of spatial decision-support systems.

A UML EQUIVALENT FOR FIELDS

Despite the success of object-oriented data modeling, its fundamental assumption that the geographic world is populated by *things* that can be grouped into classes ultimately limits its application, and creates a distortion — a sense of square peg in round hole — when used to structure geographic information. Many geographic phenomena are continuous, and the task of breaking them into discrete things limits the questions that can be asked about them, and the applications that can be built on databases. For example, terrain is continuous, and breaking it into discrete triangles inevitably creates distortion. Similarly, rivers and roads are continuous, and must be broken into pieces at nodes in order to fit the concepts of object orientation. Unfortunately, GIS software does not record the lineage of objects that are used to represent fields, and as a result is capable of irrational acts. For example, if terrain is represented by a collection of digitized isolines, it is possible to edit their positions so that they cross, even though this is impossible in reality. Similarly some systems allow polygons representing a variable such as *owner* to be moved around freely, violating the requirement of any field that each location in the plane map to exactly one value of the function.

We believe that a first step in correcting this deficiency would be through the definition of an equivalent of UML for fields. If the isolines of a terrain representation were identified in this way, then methods could be associated that would prevent intersection during editing. In Figure 2.6 we present one possibility, in which the class of an object-oriented design is replaced by the discretization of a field. For example, in a raster GIS with each layer exactly coincident in space, we would have a single discretization (the raster) with a series of variables, each corresponding to one of the layers and representing the fields that are discretized using this raster. The details of the discretization — spatial resolution, geo-registration, compression method, and so on — would be defined as a property of the discretization itself.

In UML several types of relationships are recognized between classes: inheritance, association, aggregation, and composition. One type of relationship between boxes in this field-based schema might represent the techniques needed to change representation, and might be termed *transformation* relationships. They might include spatial interpolation, to transform between point samples and a TIN, for example, or digitized isolines and a regular sampling grid, or the resampling or aggregation methods needed to change a raster's spatial resolution. In the equivalent of an inheritance relationship, we would expect each box to specialize one of the six types of discretization commonly found in GIS, by defining the exact details of the discretization's geometry. Some discretizations, such as the state boundaries of the United States, would have a very large number of variables, while in other cases, such as a TIN representation of terrain, only one variable would likely be associated with the box. Some discretizations, such as the state boundaries, might apply to a temporal sequence of snapshots, while others, such as the isobars of a weather map, would necessarily change with every snapshot.

In the figure, which shows a simple example of this graphic way of representing fields, the seven boxes across the top represent seven types of spatial discretization (the normal six of GIS, plus the finite-element mesh). The names are shown in

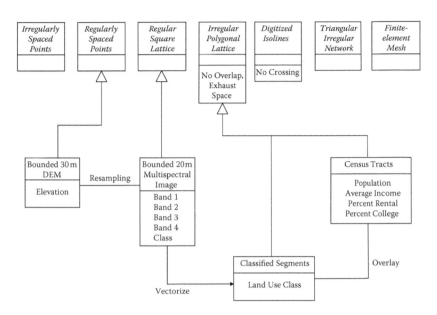

FIGURE 2.6 A possible graphic representation of continuous fields. Each box represents a distinct spatial discretization, with associated variables. Open arrows indicate inheritance or specialization relationships. Other connections indicate methods for transforming between discretizations (see text for further details).

italic, following the normal UML convention denoting an abstract class. Two types of discretization are shown with associated methods: irregular polygons representing a field must not overlap and must exhaust the space (*planar enforcement*), while polylines representing the isolines of a field must not cross. Open arrows indicate inheritance, and in this case two specific rasters are shown, one a DEM with a 30-meter spacing between sample points, and one an image, each pixel of which carries the values of four bands and a classification. Two specific discretizations specialize the irregular polygon type, one the result of vectorizing the classified image, and one recording several variables collected for census-tract reporting zones. Other links between boxes represent methods of transformation between discretizations, and may be directional if the transformation is appropriate only in one direction.

CONTINUOUS VERSUS ONE-TIME ANALYSIS

One of the corollaries of a largely static view of the world is that analysis can proceed at a leisurely pace, since the data will not change before it is completed. GIS has largely adopted what might be termed a *project-based* or *one-time* approach, in which data are collected, analysis is conducted, and results are presented and published over a fairly lengthy period of time. But as geographic dynamics become more and more central to GIScience, the fact that data change continuously in a dynamic world forces us to rethink this basic aspect of the paradigm. Applications such as wildfire management (http://activefiremaps.fs.fed.us), early warning of famine (www.fews.net), and response to emergencies all dictate a pace of analysis that matches or exceeds the pace of actual change. In many cases the need is for a continuous monitoring, in which analysis constantly responds to new data.

In this context it is interesting to note the paradigms represented by the leading GIS software products. Intergraph's GeoMedia (www.intergraph.com) could be considered to have a pipelike structure, with data at one end and the user's screen at the other. Analysis is conceptualized in the form of filters that are interposed on the pipe, allowing the user to expose different views of the data, or to perform simple statistical manipulation. This architecture is clearly much more compatible with the notion of a dynamic database than the traditional one.

SENSOR NETWORKS

Recently there has been much interest in the concept of sensor networks, or distributed collections of interconnected sensors that transmit measures of their environment, along with information about location, to central servers that interpret, compile, and redistribute data to users. The U.S. National Science Foundation has funded a sensor-network graduate program at the University of Maine and a Science and Technology Center at the University of California, Los Angeles, among other projects.

It is important to recognize that the sensors may range from inert, fixed objects to GPS-enabled devices carried by humans, to humans relying on the normal senses. In this sense the term *citizen science* is relevant, describing as it does the use of extensive networks of human observers to collect and compile useful data. The Christmas Bird Count (www.audubon.org/bird/cbc) is only one example of an increasing number of ways in which individuals empowered by mobile technologies become effective sensors of useful geographic information. Recently, a number of projects have built on the success of Wikipedia (www.wikipedia.org) by encouraging individuals to uplink geographic information about their local areas, particularly information that can be used to enrich the *gazetteer* or names layer. Other groups are developing maps by driving streets in vehicles equipped with GPS. In each of these cases groups of individuals provide a cost-effective alternative to the traditional mapping agencies, and a way of addressing their problems over declining budgets and increasing demands.

Sensor networks and citizen science offer interesting ways of addressing the supply of data about geographic dynamics. But many questions arise: How can masses of potentially conflicting data be assembled into useful databases? How can quality be assured, and what kinds of institutional arrangements would be needed? And what strategies can overcome the scaling issues of massive networks?

CONCLUSIONS

The first section of this chapter addressed the question of the domain of geographic dynamics, and concluded that it included virtually all disciplines that deal with the surface and near-surface of the earth, and virtually all mechanisms that modify and transform that domain. This is an enormous charge, requiring effective communication and collaboration between a highly distributed set of researchers and users. Its intersection with the discipline of geography is uneven, since it encompasses some areas where geographers have made substantial contributions, including hydrology and biogeography, and others, such as tidal dynamics, where it would be very hard to find a specialist geographer.

Within this domain, we argued that the focus of GIScience should be on the generic — on the tools, data models, software, standards, and other structures that can support the domain as a whole. This is a difficult task, requiring a comprehensive knowledge of the domain and an ability to generalize about its requirements. But such generic support can be enormously cost-effective, interoperable, and helpful in integrating a multidisciplinary enterprise.

To provide a conceptual framework, we invoked the concepts of continuous fields and discrete objects, arguing that all processes are defined as interactions between fields, interactions between objects, or interactions between objects and fields. Within this framework it was possible to review existing tools and other forms of support, and to identify significant gaps where little or no generic support exists. We argued that the distinction between nomothetic and idiographic knowledge was also relevant, in that knowledge of form, the traditional focus of GIScience, belonged to the idiographic realm while knowledge of process was essentially nomothetic. We also argued that the structure of modern GIS, with its separation between the local detail of the database and the general procedures of the software, epitomized the idiographic/nomothetic distinction.

Finally, we identified five areas, or gaps, where current knowledge and technique fall far short of what is needed if GIScience is indeed to provide generic support for geographic dynamics. Some of these involve improvements to representation, some to computation, and some to the organizational frameworks in which such work is embedded. A focus by the GIScience community on these and other deficiencies in our current state of knowledge will do much to move our collective capabilities forward, and to strengthen the contribution of GIScience to the representation and computation of geographic dynamics.

ACKNOWLEDGMENTS

This research was supported by the National Science Foundation through collaborative award BCS 0417131 to Goodchild, and by ESRI through support for Glennon.

REFERENCES

Ahearn, S. and J.L.D. Smith. 2005. Modeling the interaction between humans and animals in multiple-use forests: A case study of *Panthera tigris*. In *GIS, spatial analysis, and modeling*, ed. D. J. Maguire, M. Batty, and M. F. Goodchild, 387–402. Redlands, CA: ESRI Press.

Arctur, D. and M. Zeiler. 2004. *Designing geodatabases: Case studies in GIS data modeling*. Redlands, CA: ESRI Press.

Batty, M. 2005. Approaches to modeling in GIS: Spatial representation and temporal dynamics. In *GIS, spatial analysis, and modeling*, ed. D. J. Maguire, M. Batty, and M. F. Goodchild, 41–61. Redlands, CA: ESRI Press.

Bennett, D. A. 1997. A framework for the integration of geographical information systems and modelbase management. *International Journal of Geographical Information Science* 11(4):337–57.

Burrough, P. A., D. Karssenberg, and W. van Deursen. 2005. Environmental modeling with PCRaster. In *GIS, spatial analysis, and modeling*, ed. D. J. Maguire, M. Batty, and M. F. Goodchild, 333–56. Redlands, CA: ESRI Press.

Carey, G.F., ed. 1995. *Finite element modeling of environmental problems: surface and subsurface flow and transport.* Chichester, UK: Wiley.

Clarke, K. C., S. Hoppen, and L. Gaydos. 1997. A self-modifying cellular automaton model of historical urbanization in the San Francisco Bay area. *Environment and Planning B* 24:247–61.

Fotheringham, A. S. and M. E. O'Kelly. 1989. *Spatial interaction models: formulations and applications.* Boston: Kluwer.

Galton, A. and M. F. Worboys. 2005. Processes and events in dynamic geo-networks. In *Proceedings of the First International Conference on Geospatial Semantics*, GeoS 2005, ed. M. Rodriguez, I. Cruz, S. Levashkin, and M. J. Egenhofer, 45–59. Lecture Notes in Computer Science 3799. Berlin: Springer Verlag.

Goodchild, M. F. 1988. Stepping over the line: Technological constraints and the new cartography. *American Cartographer* 15:311–19.

Goodchild, M. F. 2004. GIScience, geography, form, and process. *Annals of the Association of American Geographers* 94(4):709–14.

Goodchild, M. F., M. Yuan, and T. J. Cova. 2007. Towards a general theory of geographic representation in GIS. *International Journal of Geographical Information Science* 21(3):239–60.

Kemp, K. K. 1997. Fields as a framework for integrating GIS and environmental process models. Part one: Representing spatial continuity. *Transactions in GIS* 1(3):219–34.

Kemp, K. K. 1997. Fields as a framework for integrating GIS and environmental process models. Part two: Specifying field variables. *Transactions in GIS* 1(3):235–46.

Kwan, M.-P. and J. Lee. 2004. Geovisualization of human activity patterns using 3D GIS: A time-geographic approach. In *Spatially Integrated Social Science*, ed. M. F. Goodchild and D. G. Janelle, 48–66. New York: Oxford University Press.

de Smith, M. J., M. F. Goodchild, and P. A. Longley. 2007. *Geospatial analysis: a comprehensive guide to principles, techniques and software tools.* Winchelsea: Winchelsea Press. http://www.spatialanalysisonline.com.

Sui, D. Z. 2004. Tobler's First Law of Geography: A big idea for a small world? *Annals of the Association of American Geographers* 94(2): 269–77.

Tobler, W. R. 1970. A computer movie simulating urban growth in the Detroit Region. *Economic Geography* 46:234–40.

Tomlin, C. D. 1990. *Geographic information systems and cartographic modeling.* Englewood Cliffs, NJ: Prentice Hall.

Tufte, E. R. 1983. *The visual display of quantitative information.* Cheshire, CT: Graphics Press.

Ungerer, M. J. and M. F. Goodchild. 2002. Integrating spatial data analysis and GIS: A new implementation using the Component Object Model (COM). *International Journal of Geographical Information Science* 16(1):41–54.

van Deursen, W.P.A. 1995. *Geographical information systems and dynamic models: development and application of a prototype spatial modelling language.* PhD thesis, University of Utrecht.

Vckovski, A. 1998. *Interoperable and distributed processing in GIS.* London: Taylor and Francis.

White, R., B. Straatman, and G. Engelen. 2004. Planning scenario visualization and assessment. In *Spatially Integrated Social Science*, ed. M. F. Goodchild and D. G. Janelle, 420–42. New York: Oxford University Press.

Yuan, M. 1999. Representing geographic information to enhance GIS support for complex spatiotemporal queries. *Transactions in GIS* 3(2):137–60.

Yuan, M. 2001. Representing complex geographic phenomena with both object- and field-like properties. *Cartography and Geographic Information Science* 28(2):83–96.

3 Complex Networks for Representation and Analysis of Dynamic Geographies

Steven D. Prager

CONTENTS

INTRODUCTION

Many significant advances in understanding complex geographic phenomena can be attributed to advances in representation. Issues of representation — whether related to representation for computational purposes or for visual purposes — possibly garner more attention than any other "single" topic in the GIScience literature. At the same time, however, issues of representation are almost inevitably bound to the geography of location (Batty 2005). An alternative perspective, the explicit characterization of the geography of interactions and relationships, is required to make the leap from analysis centered about where things exist to analysis addressing the consequences of changes in that existence.

In spite of advances in space-time data representation, corresponding ontologies, and the move — albeit subtle — away from strict object and field views (Peuquet 2001), additional advances are warranted. The location-centric perspective that serves as the foundation for the vast majority of GIScience research and GIS-based analysis is fundamentally limited by the constraints imposed by working in Euclidean space. The limited dimensionality of Euclidean space and the relatively strict set of entity-relationship rules imposed by even the most sophisticated spatiotemporal data models and database implementations (Marchand et al. 2004; McIntosh and Yuan 2005) still embody explicit assumptions regarding the geography of location and its role in defining the spatiotemporal process.

Moving beyond limitations in current GIS requires a perspective that moves away from a focus on structures (Batty 2005) and refocuses on explicit consideration of relationships, process, and concepts. Interestingly enough, Tobler's First Law of Geography, that everything is related to everything else but things that are nearer are more related (Tobler 1970), retains its relevance. Now, however, instead of thinking about distance in terms that are constrained by Euclidian dimensionality, we must think in terms that build on the ideas of the relationship space created by the topology of a network representation (Béra and Claramunt 2003). That is not to say that the geography of location is unimportant, but rather that the geography of location is a complement to the geography of relationships and interactions. Specifically, without the relationships formed as a function of process, geographies of location would be irrelevant. Likewise, the geographies of location give context to processes that, without location, would be meaningless.

The purpose of this chapter is to offer direction and insight regarding how network representation, complex network theory, and related concepts can be used to advance understanding of dynamic geographic domains. This chapter will explore several ideas relating to the use of networks to understand dynamic geographies. Beginning with a discussion of issues surrounding network representation, the chapter will then progress into a brief overview of existing complex network theory and the analytic opportunities that arise simply from thinking about the representation of dynamic phenomena as network data structures. The chapter will then address new opportunities for the use of networks in a dynamic geographic context and will conclude with some basic examples and a discussion of future opportunities.

NEED FOR ADVANCES IN REPRESENTATION

Complex network-based representation and analysis are underexploited in geographic information science and have the potential to offer a great deal of insight into dynamic geographic phenomena. Though networks have long played a role in geographic thought, within most modern geographic information systems network and graph structures are typically relegated to behind-the-scenes use in the areas of topology, routing, and hydrographic analysis. Haggett (1969) and Haggett and Chorely (1970) offer an early and extensive treatment of networks and graph-based analysis of geographic networks. Yet, while "description and analysis of network structures" has been characterized as a "traditional concern of geographers" (Taaffe and Gauthier 1973), advances in network-based analysis of geographic information have stalled. Recent developments in the representation of geographical concepts rely on the use of network representation (MacEachren et al. 2004), but these advanced applications are characterizing networks of concepts with limited linkages to geographic phenomena and, particularly, dynamic geographic phenomena. The area of space syntax is one example where network-based representations have been used to garner new insight into spatiotemporal issues, yet even with demonstrated integration of the theory within a GIS (Jiang and Claramunt 2002), the technique is peripheral to mainstream thinking about spatiotemporal problems.

The potential for complex network theory to increase understanding of dynamic phenomena has been demonstrated in disciplines outside of geography. Several disciplines — many in the social sciences, in fact — have embraced the idea of dynamic social networks with the theoretical underpinnings of "small-world" networks serving as a basis for explaining empirical phenomena ranging from dynamic processes, such as disease transmission (Christley et al. 2005), to the social networks formed in online learning communities (Russo and Koesten 2005). Although geographers have long embraced the concept of networks at a conceptual level (Barrett et al. 1999), the networks are largely notional, and examples that employ a large body of existing and emerging complex network theory are only just starting to appear (Grabher 2006; Grubesic and Murray 2006).

Advancements in the understanding of dynamic geographic phenomena can benefit significantly from the wealth of advances that are occurring in the area of complex network theory (Barabási 2002; Newman 2003). The crucial departure in complex network theory from the approaches described by Kansky (1963), Hagget and Chorely (1970), Taaffe and Gauthier (1973), and others has largely to do with the scale of the networks being examined. Taaffe and Gauthier (1973) describe networks as "highly complex spatial systems" and suggest that describing these networks is a highly complex task that requires substantial simplification in representation. In differentiating the study of small graphs from complex networks, Newman (2003) highlights the shift from consideration of graphs ranging from 10 to perhaps hundreds of vertices and edges to networks with potentially millions of vertices and edges. In this latter instance, it is the statistical analysis of the properties and behavior of the network as a whole that has the potential to yield unique insight into the properties of the system characterized by the network-based representation. Furthermore, whereas simple networks or graph structures have interesting measurable properties, complex networks are marked by nontrivial topological structures that emerge as a function of the scale of the network in question. These structures are often marked by properties such as heavy-tail degree distributions, high amounts of clustering, and implicit community structure, all of which have the potential to influence the behavior of a networked, dynamic system (Strogatz 2001), as well as the interaction of that system with its geographic context (Herrmann et al. 2003).

Formalization of concepts relating to models of network growth, processes taking place on networks, dynamic network attribution, dynamic network topologies, and statistical measures characterizing complex networks will all serve to increase understanding and representation of dynamic geographic phenomena. The real opportunity for advancement, however, lies with the ability of geographic thinkers to leverage network representations in ways that serve to integrate the representation of spatial relationships, conceptual or functional relationships, and dynamic variation therein. Consider, for example, changes in the global political climate. The following illustration, adapted from *The Clash of Civilizations and the Remaking of World Order* (Huntington 1996), illustrates a set of relationships that span both obviously geographic entities as well as semantically grounded concepts, such as groups of countries with similar ideology, or even a single religion.

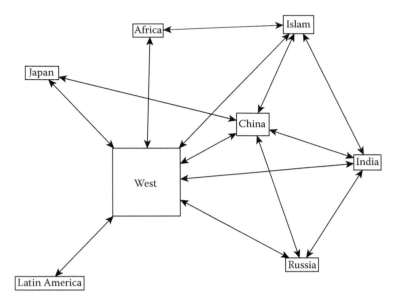

FIGURE 3.1   A dynamic network depicting the relationship between conceptual and geographic phenomena after Huntington's 1996 characterization of emerging political alignments. The size of the nodes indicates the relative centrality. In a dynamic sociopolitical context, alignments that increase centrality would tend to better position a country or ideology on the global center stage.

Huntington's notion of "emerging" suggests the inherent dynamism that is a prevalent undercurrent in the global milieu. The complexity of this problem, however, is readily illustrated by Figure 3.1, which is itself a significant simplification of the networks that actually exist. As the figure illustrates, the linkages of various geographies to the conceptual dimensions of religion have the potential to create a push-pull effect depending on the association. Though locations are fixed in geographic space (e.g., the North American continent), the idea of an emerging alignment (i.e., increasing proximity) of Africa and Islam might suggest deconstruction (distance) in the relationships between Africa and the West. While spatial-interaction theory supports the notion of potential and flows in a largely geographical context (Roy and Thill 2004), no such constructs exist for the functional geographies that form dynamically as a function of the association and disassociation of both geographic and aspatial concepts.

Importantly, the concept of what is a "network" is limited only by the lack of underlying, geographically centered theory. Networks may be complex representations of the same geographic space but at multiple scales, may include representations of geographic relationships based on spatial proximity but also on aspatial characteristics, and may serve to link explicitly geographic information with explicitly conceptual information. In terms of dynamism, network data structures not only support computation regarding where something exists, but also how that location may differ conceptually — statically or dynamically — from previous locations and potential future locations in both geographic space and network space.

CONCEPTUAL ELEMENTS OF COMPLEX NETWORKS

In the object view of geographic information, representation is based on the idea of objects of increasing dimensionality. Specifically, coordinates are used to specify point objects, point objects form vertices that can be connected by lines, lines can be connected to form polygons, and polygons can be extruded to form volumes. The most basic building blocks of the object-based representation of geographic entities also form the basis of network representations. In this context, networks are simply a set of vertices connected by a series of edges.

Unlike an object-based view of a set of geographic entities where location is specified by the coordinates associated with the entity in question, location on a network is relative to the vertices and edges that comprise the network. Though vertices and edges may be associated with a physical entity of any dimensionality, they may also be associated with concepts, processes, or relationships that are either completely aspatial or that transpose space and time (e.g., a telephone conversation). Thus, while elements of the network may have an obviously geographic component (e.g., a road network is explicitly geographic), the network space becomes a new defining space for analytic endeavors. This property, the ability to use network information either in conjunction with, or independent of, explicitly geographic information, offers the possibility of analyzing network characteristics, spatial characteristics, and the interaction of each with the other.

A number of network-specific properties thus merit mention. The study of complex networks originates in the study of the mathematical representation of a "graph." Graphs may have a number of characteristics that distinguish entire graphs from one another as well as properties that distinguish characteristics within the graph itself. These characteristics, in turn, facilitate the explication of complex yet systematic relationships of phenomena that are represented as networks.

One important network characteristic in the context of dynamic geographies is the idea of graph cyclicity. In many instances the networks that characterize dynamic geographies are acyclic. Specifically, things that occur in time can only be bound to things that have already occurred. One of the most studied examples of acyclic network properties is the evolving citation network of scientific literature (Jeong et al. 2003). In the case of citation networks, new articles only cite articles that already exist. An emergent property of this type of network is that the more citations an article has, the more likely that it will be cited again as the network continues to grow. In turn, the number of links increases and, commensurately, the likelihood of additional links increases in concert.

The result of preferential attachment during network growth often leads to the formation of a scale-free network (Barabási 2002). A scale-free network is one whose degree distribution — essentially a histogram of the number of edges connected to each vertex — follows a power-law. That numerous real-world networks follow power-law distributions (e.g., the World Wide Web, certain social networks, etc.) and that many do not (e.g., electrical grids and railway networks tend to have exponential distributions) (Newman 2003) provides a benchmark for measuring and understanding the potential for certain systematic behaviors in the networks formed by dynamic geographies. Strogatz (2001), for example, highlights the idea of a cascading power failure as an example of a dynamic process occurring across a network. The structure of the power network results in certain properties that could serve to enhance or diminish the resilience of the network to system-wide collapse.

COMPLEX NETWORKS IN A GEOGRAPHIC CONTEXT

Batty (2005) proposes two related models that support the use of networks for representing dynamic geographic phenomena. The first model is premised on the idea that the number of objects in the network remains fixed, but the connections between objects are subject to change. The dynamism in this case is a function of the capacity for connections to develop or dissolve. In essence, this is a model of dynamic connectivity wherein existing entities are coupled or uncoupled by an existing process. Batty's second model is the idea that both new objects and new links are introduced (this is more in keeping with the notion of the scale-free network mentioned above, and this is the more general case of the two models). In this model, the dynamism is a function of the two sets of processes, the processes that lead to the generation of new entities as well as the process that lead to the connection or dissolution of entity-entity relationships.

In contrast to networks wherein objects are fixed and the connectivity between them varies, networks that expand and recede are capable of supporting a significantly more complex set of dynamic behaviors. When considering the idea of network growth, one might posit that the distance relationships in networked systems offer an obvious example of the applicability of Tobler's First Law to a non-Euclidean geo-space (Miller 2004). What makes a network operational, then, is the set of rules that serve to define the basis for connectivity. Though not specifically addressing dynamic geographies, the notion of relative adjacency (Béra and Claramunt 2003) serves to identify a network that is the complement, more specifically the dual, to a boundary-based representation of countries. The network-based representation coupled with the relative adjacency operator facilitates the visualization and analysis of the countries based on semantic relationships as well as the physical relationships of the countries in question. If the basis for the relative adjacency operator were subject to change over time, the resultant network would also change over time.

From a dynamic perspective, the notion of network construction via semantic or thematic association opens the doors to several analytic possibilities that are atypical of modern geographic information systems. The example of network-based representation of various property transactions over time (Sriti et al. 2005) is a good example. Sriti et al. illustrate how property location, changes in ownership, and property subdivision create an evolving spatiotemporal network that, when viewed from multiple perspectives (e.g., space, time, ownership), offers insight regarding important geographic locations, critical changes in ownership, and critical junctures in time. In effect, the series of parcel transactions become a self-organizing network with explicit characteristics in the dimensions of space, time, and ownership. That networks can be used to simultaneously characterize several dimensions of the property transactions offers new analytic opportunities significantly out of scope of a traditional geographic information system.

DYNAMIC NON-EUCLIDEAN SPACES

The previous section illustrates the notion that networks can be used to represent dynamic relationships between existing and emerging geographic phenomena. Semantic association leads to the reaffirmation of an important geographic notion, the idea of a conceptual space (Gatrell 1983). Conceptual spaces are spaces that exist as a function of distance relationships that may or may not have a Euclidian metric as their basis. When conceptual spaces are overlaid with geographic space, a functional geography emerges: That is, the geographic location has the potential to play a role in influencing the basis for the non-Euclidian metric (see Tobler 1976). In differentiating the notion of properties and concepts, Gärdenfors (2000) suggests that the properties and concepts that we use "to come to grips with the surrounding world" are dynamic in that we are constantly learning new concepts and adjusting old concepts in light of new experiences. Arguably, the reciprocal influences of network structure and geographic space on one another are similarly dynamic and result at least partially from previous "experiences." Changes occurring in geographic space may consequently affect the non-Euclidian characteristics of the network, and changes in the network certainly have the potential to exert influence on geographic space. In a network context, a functional geography is thus one that exists in geographic space but occurs relative to a nonplanar network space as a function of the relationships between various networked entities. Transportation-infrastructure networks serve to relate multiple geographic locations as a function of the connectivity between those locations, and a new set of functional geographies forms based on the characteristics of and basis for infrastructure connectivity.

Empirical data on domestic air travel help illustrate the relationship between network and geographic spaces. The following example is based on airline flight data compiled by the United States Bureau of Transportation Statistics. The database used in this analysis, entitled "Airline Origin and Destination Survey (DB1B)," is a sample of approximately 10 percent of all the flight coupons issued. A flight coupon is simply a document detailing the origin and destination of flights within any given passenger itinerary. Beginning with the first quarter of 1993, the first quarter of each of the 13 years included in this study was used to create the network characterizing air-traffic configuration for that year. It is important to note that 1993 is an interesting year in the history of airlines, as the first regional jet service was introduced in 1992 (Wong et al. 2005). Even with the use of only one quarter of the annual data, the total record count for the 13-year period included over 70 million flights.

Origin and destination data were used to construct network representations for each of the 13 years. Networks are typically thought of in graph-theoretic terms $G=(V,E)$ where graph G is comprised of a finite set of V vertices connected by a finite set of E edges. Each origin/destination pair was added to an empty network data structure, and edge values between connected nodes were calculated by accumulating the total number of connections between each origin and destination pair. For purposes of this analysis, the network was assumed to be nondirected (i.e., travel from A to B was equivalent to travel from B to A) and cyclic (i.e., connections within any given year were considered temporally independent).

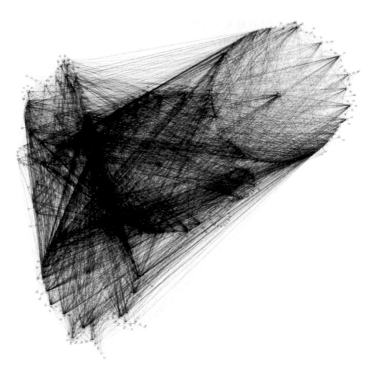

FIGURE 3.2   The network generated by airline travel in the first quarter of 2005. Implicit network structures (e.g., hubs) are evidenced by the layout algorithm that attempts to take relative connectivity and edge weight into account.

One of the challenges in working with the systems of complex networks is that the relationships formed as a function of the networks in question are often very difficult — if not impossible — to visualize. As Figures 3.2 and 3.3 illustrate, visual inspection offers only a clue regarding the structure and properties of the network in

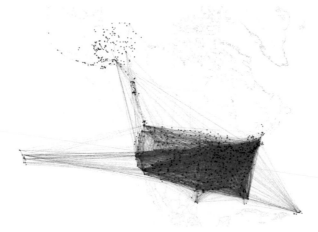
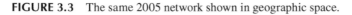

FIGURE 3.3   The same 2005 network shown in geographic space.

question. A challenge, then, is to examine the networks in such a way that meaningful information can be gathered from the network representation. This challenge is compounded when one considers that the network is dynamic and changing continuously through time. With the airline network, such changes include the addition and subtraction of different connections between airports, the addition or removal of airports altogether, and even the change in the numbers of flights between persistently connected nodes. The result is a dynamic, multidimensional representation of connectivity that, if properly examined, can offer insight regarding the interaction of a complex system with geographic space.

One of the most important unifying principles regarding geographically embedded networks is likely geography itself. Networks of geographic phenomena are fundamentally spatial and, consequently, space can be used to organize information and to provide an analytic context for examining network characteristics. For this reason, the network is constructed such that the edge weights (i.e., the values associated with the edges connecting any given origin/destination pair) are a function of the reciprocal of the total number of flights between two airports rather than geographic distance. Since geographic distance is both implicit and static (e.g., St. Louis and Seattle are always 2,751 kilometers apart), the reciprocal of total passenger flights offers another measure of how "close" two locations are — the more passengers, the larger the denominator of the edge weight and the more related the two locations are as a function of the people traveling between them. Again, this is in keeping with Gatrell (1983) and Gärdenfors (2000) and the idea of a conceptual distance — the network facilitates the characterization of a set of dynamic relationships that are complementary to geographic space but distinct in their representation. Whereas the geographic distance between two locations is fixed, the functional distances are subject to varying considerably, especially in relation to the temporal scale in which the network is observed.

In that the generated representation of domestic air transportation has a geographic complement, certain characteristics of the network may be visualized spatially to offer a new perspective on the dynamics of the structure and function of the network. One important characteristic of complex networks is degree, the number of edges connecting to a single node on the network. Degree forms the basis for the aforementioned emergence of scale-free networks (Newman 2003). Given that degree is associated with individual nodes, and nodes in this example are associated with fixed geographic locations, the spatial characteristics of degree distribution may be examined both quantitatively and visually within a GIS.

From a geographic perspective, the distribution of degree is indicative of the role that any particular geographic region has in sustaining connectivity of the network at any given point in time. Figure 3.4 utilizes kernel density estimation to spatialize the degree of the airline network connectivity. As is illustrated, that connectivity is subject to variations over time.* From a dynamic perspective, the relocation of

---

* As mentioned, the database used to create the networks in this study is comprised of the first-quarter data extracted from the Airline Origin and Destination Survey (DB1B) and is a 10 percent sample of airline tickets from reporting carriers. It is possible that some of the variation in the characteristics of the resultant networks is a function of the sampling methodology employed by the Bureau of Transportation Statistics, but the extent of any effect in this regard is unknown.

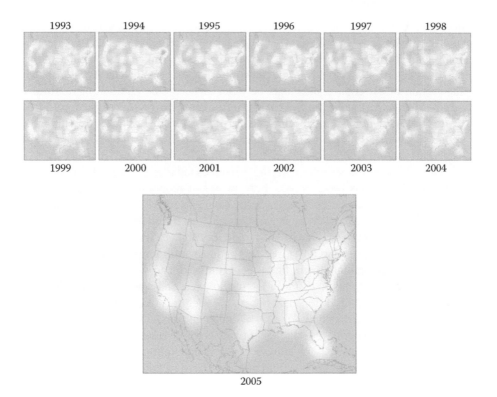

FIGURE 3.4   Change in the regional distribution of number of connections for individual nodes over time.

centers with high degree is indicative of how the structure of the network changes in response to economic factors and business trends. Examination of the illustrated time series seems to indicate a gradual consolidation of high levels of connectivity to an increasingly fewer number of regions. In turn, since high connectivity tends to be concentrated in fewer regions, if regional traffic is disrupted, the whole system tends to be more adversely affected. This relationship, the relationship between geographic characteristics and network resilience, is an important area for additional research.

The network conceptualization of distance facilitates the calculation of several additional metrics that can serve to characterize the dynamics of the network over time. As mentioned, the geographic distance between nodes on a network is a static phenomenon. At the same time, however, one aspect of the association between different geographic regions is the number of people traveling between those regions. This is in keeping with Miller (2004), who suggests that, while nearness is the "central organizing principle" of geographic space, it is not required to be a function of a Euclidian metric. Thus the relationship between places in space is not only a consideration of how far apart they are in Euclidean terms, but also in the semantic basis for an association. The severe acute respiratory syndrome (SARS) epidemic of 2003 illustrates the manner in which the airline network increased proximities of

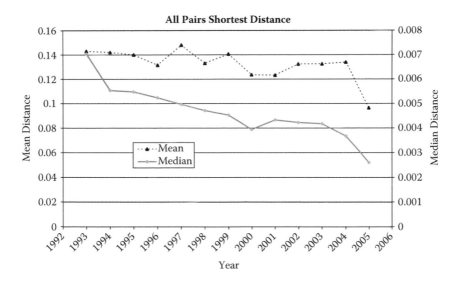

FIGURE 3.5   Utilizing all pairs shortest distance for a general characterization of connectivity as a function of network structure. In this context, network distance is the reciprocal of the number of connections between two nodes — the smaller the number, the more closely connected the two nodes.

seemingly distant locations, resulting in the rapid spread of the disease (Bowen and Laroe 2006).

In examining the spatiotemporal movement of SARS through the airline network, Bowen and Laroe (2006) observe that the initial rapid spread of the disease was attributable to the high levels of accessibility associated with certain portions of the network. Accessibility in the air-transportation network is largely defined by the hub and spoke structure that characterizes such networks (O'Kelly and Miller 1994). As such, the consequences of the function of the hub in disease transmission must also be considered in relation to the geographic location of the hub. Another important consideration, however, is that the network itself is subject to change over time. Figure 3.5 illustrates two trends in network connectivity for U.S. domestic travel as a function of network distance. As illustrated, both the mean and median network distance of the calculated all pairs shortest paths between geographic locations is trending downward. From these trends we can infer that, over time, overall network connectivity is both increasing and increasingly dependent on linkages between "close" nodes (i.e., nodes that are strongly connected by a large number of flights). These observations are in keeping with those of Wong et al. (2005), who note that even in the presence of regional jet augmentation of the airline infrastructure, airline hubs retain their role as concentrators of air-travel activities. Given the apparent increasing dependence of the network on a hub-centric topology, the relationship between the hubs and their geographic locations merits special consideration.

NETWORK-BASED FUNCTIONAL GEOGRAPHIES

The concentration of connectivity has direct bearing on the functionality, efficiency, and vulnerability of the network. In an examination of the neuronal connectivity of the brain, Sporns and Tononi (2002) observe a variety of network patterns that are indicative of functional patterns of connectivity associated with specific perceptual and cognitive states. This functional connectivity occurs in the context of the broader anatomical network (the set of neurons and the corresponding synaptic connections at a given point in time) and suggests that different portions of the anatomical neuronal network are associated with the different functions of the brain. This is, arguably, similar to a transportation network in that different nodes on the network are functioning in several ways. In the case of the U.S. domestic airline travel networks in this study, the hubs serve as both origins and destinations, but also as bridges between otherwise disparate geographies (particularly international destinations in some instances). The result is that, as a function of their position in network space, different geographies take on different functional roles (e.g., different hubs connect different regions, certain hubs may dominate accessibility during certain times of the year, etc.). In that the network is dynamic and subject to change in response to both endogenous and exogenous factors, the functional geographies are consequently dynamic as well.

The connectivity of numerous real-world networks, including transportation networks, is often a function of the scale-free nature of the network in question (Zhao et al. 2005). While the scale-free configuration tends to ensure resilience to random network interdiction, a consequence of the hub-dependent connectivity is fragility in the event of the loss of the most connected nodes (Huang et al. 2006). Consider the example of Internet and telecommunications backbones addressed by Grubesic and Murray (2006). Grubesic and Murray observe that geography, specifically distance, tends to influence the formation of telecommunications networks in such a manner that their development follows something of a hub and spoke model. If a node were removed and that node were serving as a highly connected hub, then the disruption to network accessibility would have the potential to be significant (i.e., some portions of the network could become completely disconnected, while the remainder of the network would need to handle increased load). The removal of a node from a telecommunications network is somewhat analogous to the removal of a node from a transportation network. If a secondary airport is removed from the network, the overall effect on the functionality of the network is minimal. On the other hand, if a major hub is removed (e.g., due to severe weather as happens in many locations every year), the consequences can ripple throughout the entire system. Given the increasing dependence on hubs to sustain and increase connectivity over time (Figure 3.5), there is a commensurate increase in the impact on the viability of the network with the removal of any given hub.

An interesting dichotomy arises when one examines the configuration of network space relative to the configuration of the underlying geography. In the geographic space of the continental United States it would seem that, perhaps, the geographic center would tend to roughly correspond with the center of the corresponding transportation network. The network structure facilitates the calculation of measures of centrality (again, based on all pairs traversal of the network) and, interestingly, for several of the years in this study, Anchorage, Alaska, is the most central node on the network. This finding is echoed in a similar analysis by Guimera et al. (2005), who observe that Anchorage is one of the most central nodes in the global airline network, second only to Paris. In both cases, this raises the dilemma as to why a relatively unconnected node would have such a central place in either of the networks in question. Specifically, why would a geographic location that is relatively distant from many of the other geographic locations in the network have a central position in network space? Guimera et al. (2005) recognize that, due to geography, most Alaskan airports are connected principally to other Alaskan airports. Geographically, Anchorage serves as a consolidator of the many distributed communities and, in turn, as the connection for those communities to the remainder of the network and the continental United States. The "anomalous centrality" (Guimera et al. 2005) of Anchorage stems from the function of its geographic location in linking two parts of the nation, but the nature anomaly is not evident without inspection of both geographic and network spaces.

From a dynamic perspective, the geographic location that supports a particular network function is subject to change as the network structure changes. In a simulation of vertex and edge removal "attacks" on a network, Holme et al. (2002) suggest that fundamental measurements of network structure are subject to change as important edges are removed. In that measures of connectivity and centrality are subject to change as a function of changes to the network, the geographic locations that are playing specific roles with regard to network functionality are also subject to change. This is a very important dynamic geographic phenomenon that merits substantial exploration in a variety of domains.

DYNAMICS IN AND ACROSS NETWORKS

Though a single, minimally attributed, connection-based generative rule was used to create the functional geographic network describing airline travel, that network is by no means the only possible representation of the domestic air-transportation system. For example, the networks created by airline travel could be linked to the geographic areas they connect instead of the actual airports (e.g., Newark, LaGuardia, and John F. Kennedy airports all serve New York City), as in the analysis of Guimera et al. (2005). Similarly, distance between nodes in the network space could be simultaneously defined by geographic distance, percentage of on-time flights, or any number of other quantities. Furthermore, completely different networks that share similar nodes (e.g., highways between cities) could also be linked via the nodes in common. Thus, the entities serving as nodes in the network can be simultaneously incorporated into multiple network representations. In turn, these shared entities link different networks, each with potentially different spatiotemporal characteristics and, hence, different functional geographies. In the context of functional geographies, the shared entities that serve to link two or more networks can serve as the focal point for analysis that would have been impossible with any other type of representation. Recall the notion of the scale-free network and the commensurate ideas of preferential attachment. If a particular city served as a "hub" within the network formed by the airline transportation network *and* that city was located in a similar hub in the ground-transportation network, then the possibility exists for disruptions in one network space to move across network spaces and cause disruption in another network space.

The movement of a phenomenon through network space and, consequently, geographic space is known as percolation and is supported by underlying percolation theory. Frequently used in the context of understanding disease epidemiology (Moore and Newman 2000), the robustness or fragility of the Internet (Callaway, Newman et al. 2000), and even characteristics of species distribution (Brooks 2006), percolation theory is one basis for understanding how a network supports or inhibits propagation of phenomena through the network. Batty (2005) points out that gradual increases in connectivity can lead to sudden increases in percolation as the network becomes more connected. Given the possibility that one network might serve to induce connectivity in another spatially coincident network, however, functional geographies must be considered across multiple networks to get a true picture of connectivity. For example, ground-transportation connectivity could "bridge" two midsize, relatively close cities straddling an international border that do not otherwise have connectivity in the airline network. In the case of percolation of disease, information, or other phenomena, the network bridge induced by the proximal geography and connectivity as a function of the road network could significantly increase rates of percolation through the airline network.

The premise that a network between objects serves as a conduit of flow offers a number of opportunities for better understanding the dynamics occurring as a function of the presence of network connectivity. In the context of the network as an enabler, it is important to realize that the complexity of the network is not only a function of the structure of the network but also the interactions occurring across that structure. In the study presented here, though the structure of the network varies over the course of the time span of this study, the variation in the interactions occurring across that same network is substantially more significant. Keeping in mind that the complexity and connectivity of networks occurs in response to both exogenous and endogenous factors, consideration of such factors is vital for a complete understanding of the role of network connectivity and network dynamics in relation to both geographic and network-specific phenomena.

CONCLUSION

In the context of the analysis of dynamic geographic domains, the concept of what is a "network" is limited only by the lack of underlying, geographically centered theory regarding the relationships that form network and network-based interactions. Networks may be complex representations of the same geographic space but at multiple scales (Brooks 2006), may include representations of relationships based on proximity but also on aspatial characteristics (Béra and Claramunt 2003), and may serve to link explicitly geographic information with explicitly conceptual information (Sriti et al. 2005).

One intent of this chapter is to illustrate not only the validity of using network-based constructs to better understand dynamic geographic domains, but also to demonstrate that the integration of network approaches is not simply the horizontal transposition of one knowledge domain into the geographic domain. This chapter illustrates that increased understanding of geographic space using networks can, and should, occur based on contributions from physics (Barabási 2002; Newman 2003), psychology (Sporns and Tononi 2002), ecology (Brooks 2006), and geography (Grubesic and Murray 2006). Network-based understanding of dynamic geographic domains should evolve within the context of several complementary theoretical areas. Specifically, work in the area of ontologies (Agarwal 2005) is particularly relevant, as the basis for defining network generative rules should be based on sound understanding of geographic domains. In that substantial energy has been invested in issues relating to geographic and spatiotemporal representation (Yuan 1999; Hornsby and Egenhofer 2000; Peuquet 2001), new endeavors in network space should consider opportunities and gaps presented in earlier works.

Immediate opportunities exist to employ complex network theory to extend research in geography and, particularly, dynamic geographic domains. In the case of the air-transportation network and numerous other technical networks with a geographical complement, the evolution of the network itself is only one aspect of the dynamic nature of the system. Perhaps more important is that the network is a conduit to a large number of dynamic transactions or flows that occur along network edges. At any given moment in time and space, the precise structure of the network is determined by the instantaneous set of connections about to be completed (e.g., airplanes arriving at their destinations) and the set of potential connections about to be initiated (e.g., airplanes about to depart their origin). There is great need to understand how best to optimize the function of the network given exigent circumstances ranging from weather-related problems to terrorist interdiction. Any response to such a scenario would need to occur in the context of the evolutionary state of the network as well as the instantaneous status of the network at the moment of disruption. The importance of the interplay of geography and the functional spaces generated by the network requires explicit consideration in such a context, for the characteristics of the functional space are required to rapidly change from a configuration associated with the evolutionary structure of the network to one that emerges based on the nature of the instantaneous state change.

At a more fundamental level is the need for additional formalization of the relationship between geographic theory and complex network theory. Strogatz (2001), Hermann (2003), and Newman and Girvan (2004) all illustrate dynamic properties of complex networks that have the potential to significantly inform analysis of dynamic geographic domains. While Batty (2005), Grubesic and Murray (2006), and others have demonstrated the potential for integrating complex network theory and geography, clear articulation regarding the interrelationship between dynamic geographic domains and complex network theory is required. Inroads have been made in the area of visualization (Ruggles and Armstrong 1997; Fabrikant et al. 2004), but a number of challenges with regard to representation and analysis of would-be network datasets remain (Tobler 1987). This chapter illustrates that some properties of networks have natural geographic complements, while other properties are more a function of the nontrivial structures of the network in question — the extent to which these characteristics can be formalized should lay the foundation for a wide array of new analytic endeavors.

Once the potential relationships between geographic theory and complex network theory are further clarified, there exists a great deal of potential for directly integrating related analysis into existing geographic information systems. Network structures afford an alternative view to the map space that is prevalent in typical geographic information systems, yet many of the objects used to create nodes on networks are tied to geographic locations and represented in the GIS. Network-based analyses have the potential to shift the focus of GIS from representation constrained by location to representation with the innate capacity for supporting dynamic constructs. This shift, from the geography of location to the geography of relationship and interaction, has the potential of opening many new doors of inquiry and, ultimately, increasing our understanding of spatiotemporal phenomena and dynamic geographic domains.

REFERENCES

Agarwal, P. 2005. Ontological considerations in GIScience. *International Journal of Geographical Information Science* 19(5):501.

Barabási, A.-L. 2002. *Linked: The New Science of Networks*. Cambridge, MA: Perseus.

Barrett, H. R., B. W. Ilbery, A. W. Brown, and T. Binns. 1999. Globalization and the changing networks of food supply: The importation of fresh horticultural produce from Kenya into the UK. *Transactions of the Institute of British Geographers* 24(2):159–74.

Batty, M. 2005. Network geography: Relations, interactions, scaling and spatial processes in GIS. *Re-presenting GIS*. P. Fisher and D. Unwin. West Sussex, England: John Wiley & Sons:149–69.

Béra, R. and C. Claramunt. 2003. Topology-based proximities in spatial systems. *Journal of Geographical Systems* 5(4):353–79.

Bowen, Jr., J. T. and C. Laroe. 2006. Airline networks and the international diffusion of severe acute respiratory syndrome (SARS). *Geographical Journal* 172(2):130–44.

Brooks, C. P. 2006. Quantifying population substructure: extending the graph-theoretic approach. *Ecology* 87(4):864–72.

Callaway, D. S., M.E.J. Newman, S. H. Strogatz, and D. J. Watts. 2000. Network robustness and fragility: Percolation on random graphs. *Physical Review Letters* 85(25):5468.

Christley, R. M., G. L. Pinchbeck, R. G. Bowers, D. Clancy, N. P. French, R. Bennett, and J. Turner. 2005. Infection in social networks: Using network analysis to identify high-risk individuals. *American Journal of Epidemiology* 162(10):1024–31.

Fabrikant, S. I., D. R. Montello, M. Ruocco, and R. S. Middleton. 2004. The distance-similarity metaphor in network-display spatializations. *Cartography and Geographic Information Science* 31:237–52.

Gärdenfors, P. 2000. *Conceptual space: The geometry of thought*. Cambridge, MA: MIT Press.

Gatrell, A. C. 1983. *Distance and Space: A Geographical Perspective*. Oxford, England: Clarendon Press.

Grabher, G. 2006. Trading routes, bypasses, and risky intersections: Mapping the travels of "networks" between economic sociology and economic geography. *Progress in Human Geography* 30(2):163–89.

Grubesic, T. H. and A. T. Murray 2006. Vital nodes, interconnected infrastructures, and the geographies of network survivability. *Annals of the Association of American Geographers* 96(1):64–83.

Guimera, R., S. Mossa, A. Turtschi, and L. A. N. Amaral. 2005. The worldwide air transportation network: Anomalous centrality, community structure, and cities' global roles. *PNAS* 102(22):7794–99.

Haggett, P. 1969. Network models in geography. *Integrated Models in Geography; Part IV.* R. J. Chorley and P. Haggett. London, Methuen:609–68.

Haggett, P. and R. J. Chorley (1970). *Network Analysis in Geography.* New York: St. Martin's Press.

Herrmann, C., M. Barthelemy, and P. Provero. 2003. Connectivity distribution of spatial networks. *Physical Review E (Statistical, Nonlinear, and Soft Matter Physics)* 68(2):026128-6.

Holme, P., B. J. Kim, C. N. Yoon, and S. K. Han. 2002. Attack vulnerability of complex networks. *Physical Review E* 65(5):056109.

Hornsby, K. and M. J. Egenhofer 2000. Identity-based change: A foundation for spatio-temporal knowledge representation. *International Journal of Geographical Information Science* 14(3):207.

Huang, L., L. Yang, and K. Yang. 2006. Geographical effects on cascading breakdowns of scale-free networks. *Physical Review E (Statistical, Nonlinear, and Soft Matter Physics)* 73(3):036102-4.

Huntington, S. P. 1996. *The Clash of Civilizations and the Remaking of World Order.* New York: Simon and Schuster.

Jeong, H., Z. Néda, and A.-L. Barabási. 2003. Measuring preferential attachment in evolving networks. *Europhysics Letters* 61(4):567–72.

Jiang, B. and C. Claramunt. 2002. Integration of Space Syntax into GIS: New Perspectives for Urban Morphology. *Transactions in GIS* 6(3):295.

Kansky, K. J. 1963. Structure of transportation networks: Relationships between network geometry and regional characteristics. *Department of Geography.* PhD Diss. Chicago: University of Chicago. 155.

MacEachren, A. M., M. Gahegan, and W. Pike. 2004. Visualization for constructing and sharing geo-scientific concepts. *Proceedings of the National Academy of Sciences of the United States of America* 101:5279.

Marchand, P., A. Brisebois, Y. Bacdard, and G. Edwards. 2004. Implementation and evaluation of a hypercube-based method for spatiotemporal exploration and analysis. *ISPRS Journal of Photogrammetry & Remote Sensing* 59(1):6.

McIntosh, J. and M. Yuan. 2005. Assessing similarity of geographic processes and events. *Transactions in GIS* 9:223.

Miller, H. J. 2004. Tobler's First Law and spatial analysis. *Annals of the Association of American Geographers* 94(2):284.

Moore, C. and M.E.J. Newman. 2000. Epidemics and percolation in small-world networks. *Physical Review E* 61(5):5678.

Newman, M.E.J. 2003. The structure and function of complex networks. *SIAM Review* 45(2):167–256.

Newman, M.E.J. and M. Girvan. 2004. Finding and evaluating community structure in networks. *Physical Review E (Statistical, Nonlinear, and Soft Matter Physics)* 69(2):026113-15.

O'Kelly, M. E. and H. J. Miller. 1994. The hub network design problem. *Journal of Transport Geography* 2(1):31–40.

Peuquet, D. J. 2001. Making space for time: Issues in space-time data representation. *GeoInformatica* 5(1):11–32.

Roy, J. R. and J.-C. Thill. 2004. Spatial interaction modelling. *Papers in Regional Science* 83(1):339.

Ruggles, A. J. and M. P. Armstrong. 1997. Toward a conceptual framework for the cartographic visualization of network information. *Cartographica* 34(1):33.

Russo, T. C. and J. Koesten. 2005. Prestige, centrality, and learning: A social network analy-
    sis of an online class. *Communication Education* 54(3):254–61.
Sporns, O. and G. Tononi. 2002. Classes of network connectivity and dynamics. *Complexity*
    7(1):28–38.
Sriti, M., R. Thibaud, and C. Claramunt. 2005. A network-based model for representing the
    evolution of spatial structures. *4th ISPRS Workshop on Dynamic and Multidimen-
    sional GIS*, Pontypridd, Wales, UK, International Archives of Photogrammetry and
    Remote Sensing.
Strogatz, S. H. 2001. Exploring complex networks. *Nature* 410(6825):268–76.
Taaffe, E. J. and H. L. Gauthier. 1973. *Geography of transportation*. Englewood Cliffs,
    NJ.: Prentice-Hall.
Tobler, W. R. 1970. A computer movie simulating urban growth in the Detroit region. *Eco-
    nomic Geography* 46:234–40.
Tobler, W. R. 1976. Spatial interaction patterns. *Journal of Environmental Systems*
    6:271–301.
Tobler, W. R. 1987. Experiments in migration mapping by computer. *The American Cartog-
    rapher* 14(2):155–63.
Wong, D. K. Y., D. E. Pitfield, and I. M. Humpheys. 2005. The impact of regional jets on
    air service at selected US airports and markets. *Journal of Transport Geography*
    13(2):151–63.
Yuan, M. 1999. Use of a three-domain representation to enhance GIS support for complex
    spatiotemporal queries. Transactions in GIS 3(2):137.
Zhao, L., Y.-C. Lai, K. Park, and N. Ye. 2005. Onset of traffic congestion in complex networks.
    Physical Review E (Statistical, Nonlinear, and Soft Matter Physics) 71(2):026125-8.

Part II

Analysis, Computation, and Modeling

4 Exploring the Use of Gazetteers and Geocoders for the Analysis and Interpretation of a Dynamically Changing World

Daniel W. Goldberg, John P. Wilson, and Craig A. Knoblock

CONTENTS

INTRODUCTION

The observation that the world consists of a series of interesting and dynamic places is easily demonstrated by the following examples. If the region of interest were South Asia in the middle of the 19th century, an intelligence analyst focused on present-day Bangladesh might be interested in the partition of India and subsequent transformation of East Pakistan into Bangladesh, capturing what countries were involved (Burma, India, West Pakistan?) and how they have evolved over time. Hornsby (2001) used this example to (1) illustrate how the views of an object's evolution can be refined or coarsened and (2) describe a new set of temporal zoom operators to model these shifts in granularity. At the finest granularity, a data repository would need to capture all known past space-time events concerned with an entity, including relations between different entities or cross-links, such as the link between the dissolving of British India and the creation of independent India, Bangladesh, and Pakistan (Hornsby 2001). Clearly, the changes in objects and space-time events that spawn them are ongoing (witness the recent transformation of Bombay into Mumbai superimposed on a rapidly expanding metropolitan region — the latter implies that the spatial footprint requires continuous updating), and one can immediately see the great challenges incurred in the design and population of a gazetteer database that incorporates these changes. Some of these details cannot be known with certainty, indicating the need for support of uncertain queries as well.

Plewe (2002) outlined the beginnings of a theory of conceptual database design for documenting the nature of uncertainty in historical geographic information (herein termed geo-historical information). He relied on Frank's (2001, p. 670) assertion that we seem to structure our world around objects or entities, and described how these entities occur at particular places (locations), at particular times (life spans), and with particular attribute values (descriptions). These distinctions are more often fuzzy (i.e., intertwined) rather than crisp, given that multiple aspects can and are likely to interact in each manifestation. Hence, the location and description can each vary over time (within the overall time span) so that different points in space and attribute values are valid parts of the extent at various times. Plewe (2002) proposed the Uncertain Temporal Entity Model to describe the variety of causes, types, and forms of uncertainty present in geo-historical information and documented how it could be used to model uncertainty in digital stores such as the gazetteer databases discussed herein.

This discussion illustrates why the intelligence analyst may need to reconcile many different representations of the world drawn from many different sources in order to understand and anticipate a series of evolving or unfolding events and movements. The remainder of this chapter explores the latest developments of two sets of interrelated tools that help facilitate the analysis and interpretation of our dynamically changing world and the similarly dynamic data that describes it: the gazetteer and the geocoder.

MODELING THE CHANGING WORLD — THE GAZETTEER

The gazetteer, long known for its ability to translate a named geographic place into a geospatial footprint, has become the focus of a great deal of scientific research in recent times (e.g., Wang and Ge 2006, Goodchild and Hill 2007). Advances in the underlying data sources, data gathering tools, and operations necessary to derive characteristics for geographic features from the data, both spatial (e.g., its footprint) and nonspatial (e.g., its type and name) are changing the contents and uses of gazetteers. The gazetteers of old, characterized by their rigid construction, low resolution and/or accuracy, and incompleteness, have begun to give way to high-resolution, highly accurate, and complete user-specific/centric gazetteers (e.g., WikiMapia.org 2007, GeoNames.org 2007). These changes and their repercussions for the development of models capable of describing the changing world are explored in the next four subsections.

BACKGROUND

The three gazetteers most commonly used in research and practice — the ADL (Alexandria Digital Library 2007a), GNS (U.S. Board on Geographic Names 2006), and GNIS (U.S. National Geospatial-Intelligence Agency 2006) represent major milestones in the development of the gazetteer as a useful spatial-data resource but often leave those attempting to use them for high-resolution applications discouraged by the lack of detail. Undoubtedly, great effort was expended to generate these examples, first by the U.S. government to create the GNS and GNIS, and later by the University of California at Santa Barbara ADL team to create a unified gazetteer framework consisting of the GNS and GNIS as well as other data sources. Through this work, the ADL team showed that it is indeed possible to overcome many of the issues encountered integrating heterogeneous gazetteers — their publications tackle questions about ontology alignment as well as those about storage, retrieval, and the data structure of the gazetteer (Frew et al. 1998, 2000; Hill et al. 1999; Hill and Zheng 1999; Hill 2000, 2006). Understandably, the development of the ADL was focused on the integration and modeling of gazetteer data for information retrieval, and Hill's (2000) satisficing condition proved useful at the time, as resources, manpower, and time were all limited, and choices had to be made as to what should be included as gazetteer entries.

If we view the present state-of-the-art gazetteers in the broader context of GIS, we can see that gazetteers are now considered to be valuable components of spatial workflows. Perhaps most important, the three axes on which gazetteer features are based have been concretely defined; that is, that a valid gazetteer entry must consist, at a minimum, of a toponym (name), a geographic feature type, and a geographic footprint (Hill 2000). By defining the required attributes of this most basic atomic data structure, the many roles of the gazetteer could then begin to take shape. Used as a translator from textual terms to spatial footprints (Goodchild 1999), a geographic-classification system (Wang and Ge 2006), and a stand-alone spatial data model (Agouris et. al 2000), it now forms the basis for numerous geographic applications and geospatial functions spanning research communities from digital libraries (Alexandria Digital Library 2007a) to health (Dugandzic et al. 2006).

The aforementioned definition of a gazetteer continues to evolve. The current ADL Gazetteer Content Standard (GCS), for example, allows for temporal representation for the gazetteer feature itself, as well as the components of each of the axes that describe it (Alexandria Digital Library 2007b) based in part on work undertaken by the Electronic Cultural Atlas Initiative (ECAI; Buckland and Lancaster 2004) that demonstrated shortcomings in the original definition. The result of this effort is that the changing nature of geographic places can be represented across time periods in terms of changes to their names, footprints, types, and relationships in both the spatial and typographic hierarchies in which they are contained. This fundamental enhancement is responsible for enabling the gazetteer data structure to truly capture and represent the dynamism of geographic features.

Many fundamental challenges remain to be solved, however. First and foremost are the notions of completeness and accuracy (Van Rijsbergen 1979), both at the atomic (individual entry) and holistic (entire gazetteer) levels (Smith and Mann 2003, Doerr and Papagelis 2006). Atomically, completeness refers to the amount of knowledge about each axis of an entry (e.g., Are all possible names for an entry present?), while accuracy refers to the correctness of knowledge about each axis (e.g., Are all the types associated with a feature correct?). The status of these measures at the atomic level is important because they will directly impact applications that use individual features in analysis. Similarly at the holistic level, completeness (How much of the real world is represented by entries in the gazetteer?) and accuracy (How well does the descriptive data in the gazetteer represent the real world?) measures are required for evaluating individual gazetteers and for applications working with aggregate or large data sets. The challenges here are twofold: first, developing methods to assess the completeness and accuracy; and second, developing methods to actually increase both the completeness and accuracy of individual entries and the gazetteer as a whole.

A second issue is the integration of heterogeneous gazetteers. The ADL experiments proved that it is indeed possible to integrate gazetteers, and that the results are extremely valuable. They also showed, however, that integration presents an arduous, difficult, and time-consuming task, with the approach taken entirely unable to handle tens to hundreds of gazetteers. Automated means to achieve feature ontology alignment are being developed (e.g., Doerr 2001), and more will be required in order to scale to a situation where all existing field-specific gazetteers are combined into a single coherent framework. Aside from simply mapping between feature typing schemes, integration requires reasoning to determine and resolve conflicts between entries from different gazetteers. Should this be done at the integration level while the gazetteers are being merged, when integrating results sets from distributed gazetteers, or by the end user? This and other related questions will need to be addressed before very large-scale integration attempts are made, possibly using prior research from related fields (e.g., Naumann and Haeussler 2002).

A third issue relates to the handling of temporal information for geographic features. At present, it is unclear if the representation of time for geographic features is best separated from the three original axes of the gazetteer into its own, fourth axis. The current ADL GCS maintains temporal information with each axis, rightly illustrating that data about the geographic feature have a temporal extent. But we must remember that one of the main contributions of the original ADL work was the separation of the FTT structure from the gazetteer structure, explicitly indicating that while related, the two are unique and serve separate functions. The same can be said of the gazetteer and time periods (Buckland and Lancaster 2004). Due to the lack of a common temporal ontology or thesaurus, however, the easiest method is to include temporal data for all attributes as does the ADL CGS, albeit with a stub in place should these temporal thesaurii become available (Alexandria Digital Library 2007b).

In addition to general questions about the structure of time in a gazetteer, specific issues arise when considering the dynamic nature of geographic locations. Commercial buildings change tenants often (and therefore uses/types), new developments are constructed seemingly overnight, and geographic and/or administrative regions of all resolutions change names and boundaries sporadically. The temporal status attributes (i.e., current, former, and proposed) used most commonly in the ADL CGS (Alexandria Digital Library 2007b) clearly indicate the need for a more detailed representation or format. Time periods and other textual descriptions can indeed be used, but are in practice, infrequently encountered, with the exception of specific historically oriented gazetteers (e.g., Buckland and Lancaster 2004). Further questions arise, such as: (1) the circumstances when the appropriateness of either a single instant in time or time spans should be maintained; (2) whether temporal extents are/can be stated precisely or a fuzzy attribute scheme should be used; (3) how to handle overlaps in temporal extents; and (4) how to deal with uncertainty in temporal footprints.

A final, and perhaps equally contentious issue is what scale (or resolution) of geography a gazetteer should characterize, both in terms of types of features and their representations (Lam and Quattrochi 1992, Agarwal 2004). Most existing gazetteers can best be described as low resolution, with entries that encompass large geographic areas such as populated places, mountains, and so on, being by far the most common occurrence, represented as single geographic points. Higher resolution gazetteers containing a name, type, time span, and encompassing a footprint for every inch of geography in a region may be needed for some applications, however (Axelrod 2003).

The availability of new types of data and the creation of appropriate tools to exploit it to its fullest potential are changing both the contents and applications of gazetteers. The next three subsections show how emerging technologies are being used to overcome portions of the first, third, and fourth issues discussed above.

CHANGING METHODS FOR GAZETTEER CREATION

Like never before, geographic information in many forms and formats is becoming available online. Whether it is traditional GIS data sets such as raster or vector files containing geographic features, or newly emerging nontraditional data types such as online directories and phone books, information extraction tools are fast becoming robust and mature enough to reliably and accurately turn semistructured information sources (e.g., Web pages) into structured databaselike resources. While this Web-extraction technology has been around and in use for some time (see Laender et al. 2002 for a review of existing tools), only recently has it begun to be applied to the geographic domain (Himmelstein 2005, Laender et al. 2005). With these tools, scientists and researchers can now automatically harvest vast amounts of information about a geographic area of interest simply by creating "Web agents" or "wrappers"

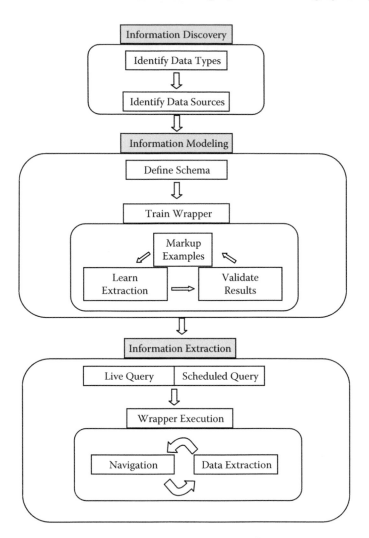

FIGURE 4.1 Typical wrapper creation and execution workflow.

around data sources that previously had no automated query ability. A typical, simplistic wrapper creation and execution workflow, including the three common stages necessary for information extraction (discovery, modeling, and extraction), are displayed in Figure 4.1 along with their component parts.

The information-discovery process begins by identifying the data types to be extracted and sources that can be used to produce them. In the modeling process, one first defines the schema for the information one wishes to extract (including the between Web page navigation structure if necessary). Through the iterative process of marking up examples of data to be extracted (training data), learning the extraction rules, and validating the results (on separate test data), the agent develops the most general and correct syntactic rules that can be applied to the pages to extract the desired information. It should be noted that validation (test) examples that fail to

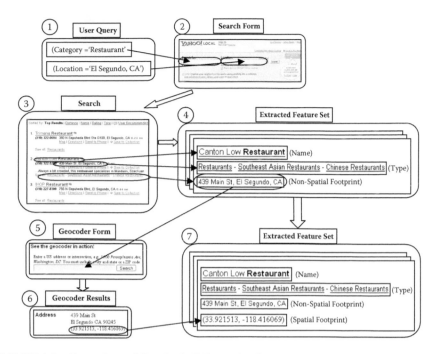

FIGURE 4.2 Example workflow for building a simple restaurant gazetteer.

extract properly given a current set of rules are typically the best choice to integrate into the training set before learning is reattempted (Knoblock et al. 2001). Having learned a sufficient set of extraction rules, an agent can be run using an execution engine (e.g., Barish and Knoblock 2005) in either an online (i.e., responding to live queries) or offline (i.e., being invoked through a scheduler) mode depending on the application and user needs. This execution engine runs each part of the agent (navigation and/or extraction) and delivers the output in a user-defined manner (e.g., by synchronously or asynchronously streaming nonpersistent XML or persistently performing database writes). It should further be noted that the description we have just provided is but a rudimentary introduction to the general idea of agent building and execution. Each component of the process is an active area of research with an expanding literature, including such topics as automatic schema identification (simplifying the markup process) and wrapper re-induction (for when the sources and their descriptions of real world features change). The interested reader should consult Laender et al. (2002) and Turmo et al. (2006).

In Figure 4.2, we display how these tools can be applied to the Web sites listed in Table 4.1 to quickly produce a simple gazetteer containing geographic features of varying subtypes of "Restaurant," a subclass of "Commercial Building," which is in turn a subclass of "Building." This workflow will require the construction of two agents, one for each of the websites in Table 4.1.

The first agent will wrap the http://local.yahoo.com (Yahoo! Inc. 2006) Web site and take as parameters the category and location of interest. Stage 1 of the workflow (denoted by the number 1 in a circle), indicates the user entering a category (i.e.,

TABLE 4.1
Data Sources Used for Creating a Gazetteer of "Restaurants"

Data Source	Parameters	Feature Information Returned
http://local.yahoo.com	Category, location	Types, names, addresses
http://geocoder.us	Address	Footprints

feature type) and a location. A first wrapper will be created around the input search form (stage 2), and the parameters supplied as agent inputs will be submitted to the site for processing. The result of this form submission will be a page containing a list of restaurants (stage 3) that will also be wrapped to extract the list of restaurants with their names, subtypes, and addresses (stage 4). These attributes contain two (toponym and type) of the three axes required to be considered a true gazetteer feature (toponym, type, and footprint). The footprint will be obtained from the second site through the process of geocoding (this process is explained in some detail later in this chapter). Obviously, the exact start and end times for when the information obtained was valid are not explicitly available, but we can associate a time period to the sources from which the data are obtained. The simplest and safest method is to assume the most conservative estimate possible, that the sources used represent the "current" state of the world and information about the features contained therein. Therefore, at this point all that is known about the temporal extents of the attributes describing these axes is that they are "current," in the parlance of the ADL GCS (Alexandria Digital Library 2007b).

A second agent will be created around the http://geocode.us site, and the first wrapper will be created around the input form located at http://geocode.us that will take an address as a parameter (stage 5). The address attribute of each of the features extracted from the http://local.yahoo.com site will be submitted to this site for processing. The resulting page will also be wrapped to extract the latitude and longitude coordinates for each address (stage 6). The spatial footprint will then be associated with the address to complete the name, type, and footprint data required for a basic gazetteer entry (stage 7).

As shown in the previous example, the ease with which these Web-extraction tools can be employed is leading to a dramatic increase in data sources for gazetteer creation and geospatial analysis. Applying these tools to the geographic domain is especially fruitful in that information typically not considered valid geographic data (e.g., phone-book data from the site http://local.yahoo.com) can be converted into data that can be mapped and interpreted geographically. As stated, the temporal footprints for the information along each of the axes may only be safely assumed "current" at each execution, but iteratively re-executing them can produce a detailed lineage of information about a feature as it exists across time periods (e.g., as tenants or uses of structures change). The ability to achieve this type of fine-scale tracking of features across time periods using spatiotemporal gazetteers has been successfully reported in research before (e.g., Agouris et al. 2000), but the advances required in the tools necessary to easily, reliably, and repeatedly generate them have until now, been missing from the literature. The types of sources described thus far provide "spotty" temporal information, and further research is required to determine what additional data can safely be assumed or derived from them beyond the information that they provide as "current."

CHANGING ROLES FOR THE GAZETTEER

This progress is enabling the rapid construction of detailed geospatial models of environments representing temporal "snapshots" (Gadia 1998) of the world as it was at one or more specific times. Depending on the geographic, temporal, or typographic interest of the party creating the gazetteer, any number of heterogeneous data sources can be wrapped and linked together using one of a number of data-integration frameworks to produce user-centric gazetteers focusing on whichever axis of the gazetteer is the most important. For example, the analysis of text to recognize and tag geographic references using natural language processing (NLP) and named entity recognition (NER) requires the use of a gazetteer that can provide an extensive list of place names for reference. For such a purpose, a gazetteer with rich name data (i.e., alternative names and spellings for places) will be a better choice than a gazetteer with sparse name data (i.e., only one name per place) (Mikheev et al. 1999, Maynard et al. 2004).

Depending on the geographic granularity of the features to be identified, different types of gazetteers can and should be created and employed. For global-scale applications, such as identifying references to countries and major cities in the *New York Times* world news section, a low-resolution gazetteer created from sources such as the *CIA World Factbook* (U.S. Central Intelligence Agency 2007) might be sufficient. In contrast, for the task of identifying geographic references in incident reports from the police blotter section of a small-town newspaper, a high-resolution gazetteer covering the local area would be required. It would be impossible for a gazetteer created using the *CIA World Factbook* (or even any well-known gazetteer presently available for that matter) to identify the geographic footprint for the text "Robbery at Starbucks on Main St," although having this more detailed information in a gazetteer format would be valuable for many purposes. In contrast, if one were to use a gazetteer created from the http://local.yahoo.com Web site in our restaurant gazetteer example, the phrase "Starbucks on Main St" could be deconstructed and associated with a geospatial footprint.

With this in mind, we will turn our attention to geospatial mediators, tools that are particularly useful in enabling the construction of user-centric gazetteers that serve these different applications. The traditional (non-geospatial) mediator architectures have long been used as a basis for information-integration projects using large numbers of heterogeneous data sources to perform complex operations (Thakkar et al. 2005). These tools excel at automatically determining the correct data sources to use for a particular problem, as well as automatically deriving an execution plan to integrate the data to achieve the desired goals. By modeling the available gazetteer data sources and the operations that take place upon those data within a mediator framework, one can leverage the heavily researched planning and execution capabilities developed in artificial intelligence for other purposes (e.g., Adali et al. 1996, Ambite and Knoblock 1997). To begin the construction of a mediator, one describes the available information sources in terms of their input parameters (known as binding parameters), output parameters, and the functions within the system that they perform (known as domain rules) along with any constraints on their scope. This set of descriptions defines the source relations for the mediator system. The binding parameters ensure that a source cannot be queried unless all required data that is needed to proceed is available, and the domain rules represent the high-level operations performed in the mediator and are used during query reformulation. The scope constraints are used to specify minimum thresholds for the sources and operations in terms of their suitability for the task(s) at hand.

Geospatial mediators (Gupta et al. 1999, Shimada and Fukui 1999) extend the functionality of traditional mediators by embedding knowledge about geography and geospatial concepts and relationships that can be used when working with geographic data in particular. For each data source and operation, a series of both spatial and nonspatial attributes can be associated and exploited as the mediator reasons about how it should proceed to solve a particular query (Raman and Hellerstein 2002). For instance, one can associate spatial attributes such as a geographic extent (e.g., a bounding box) representing the geographic coverage for which it is valid, spatial resolution, or spatial accuracy metrics (e.g., horizontal or vertical positional accuracy), and when given a choice between multiple options, the mediator can choose the data source that provides information about the area of interest (within the geographic extent) at the highest resolution and/or level of accuracy available.

Alternatively, instead of always choosing the highest-quality data sources, the mediator can be instructed to optimize one particular characteristic over another. This might be a suitable option in the case when one wishes to optimize for response time, such as during a disaster when obtaining mostly correct information quickly is more important than obtaining entirely correct information slowly. Here, if a highly accurate source's response time is lengthy, but a less, yet sufficiently, accurate source is quicker, the mediator can reason the correct execution path and choice of sources by the simple association of the response time attribute for the data sources in its domain model.

This ability gets at the heart of the process of creating a user-centric gazetteer for specific purposes. Generalizing and embedding the gazetteer-creation process into a geospatial-mediator framework, as displayed in Figure 4.3, creates the ability to tailor gazetteers to fit particular needs.

Figure 4.3 illustrates how the gazetteer-creation process can be generalized into its three component parts, each responsible for generating one of the three required axes, representing the four domain rules of the system. The first, *GetGazetteer()*, represents the concept of creating the entire gazetteer. The remaining three, *GetNames()*, *GetTypes()*, and *GetFootprints()*, are each responsible for generating the names, types, and footprints for entries, respectively. Each of these three are themselves generalized (conceptual) functions, which in turn are further broken down into a series of composite functions $f_1()$ through $f_n()$ (that can themselves be conceptual and composed of subfunctions) that work together to achieve the desired output for that axis of the gazetteer. In this manner, the designer of gazetteer-creation processes can embed the particular requirements of their application directly into the framework for the creation of the attributes for each of the gazetteer axes.

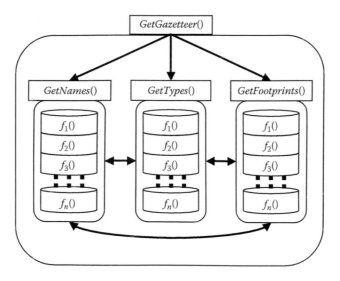

FIGURE 4.3 Generalization of gazetteer-creation process.

This type of generalization of conceptual operations just described is aptly suited for a mediator framework where the operations (domain rules) are modeled as views over the data-source models (source relations). Figures 4.4 and 4.5 ground this with an example. In Figure 4.4, the domain rules for a generalized version of the restaurant gazetteer example are provided. Here, the restriction that the feature types produced be subtypes of restaurants has been removed by decoupling the *Get-Names()* and *GetTypes()* functions from the http://local.yahoo.com source. In this simplistic example, the *GetGazetteer()* function is described as a conjunction of the three subfunctions. Parameters with dollar signs are the binding parameters that must be satisfied before execution of that function can begin, and parameters without dollar signs are return values. Thus, we see that in this example the user would first

```
GetGazetteer ($location, $type, features ):–
GetTypes ($location, $type, subTypes)^
GetNames ($location, $subTypes, typedAndNamedfeatures)^
GetFootprints ($typedAndNamedfeatures, features)
```

FIGURE 4.4 Domain rules for gazetteer generation.

```
YahooTypes ($location, $type, subTypes):–
GetTypes($location, $type, subTypes)

YahooNames($location, $subTypes,
typedAndNamedfeatures):–
GetNames ($location, $subTypes, typedAndNamedfeatures)

GeocoderUS($typedAndNamedfeatures, gazetteerFeatures):–
```

FIGURE 4.5 Source description for gazetteer generation.

enter a type parameter to get the system started, which would then cause the *Get-Types*() function to execute and gather the subtypes from the http://local.yahoo.com site (that would require the creation of another separate agent that simply extracts subtypes given a supertype). Upon the generation of the list of feature types from this function, the *GetNames*() function will query for names of features with those types within the location. Finally, once the names have been associated and the features returned, the *GetFootprints*() function will associate spatial footprints with the named features to complete the gazetteer entries. Figure 4.5 displays the source relations, identifying, for each source, which domain rules it satisfies. In satisfying a query, the mediator will generate a plan that combines the available source descriptions in the proper order to fulfill the constraints of the domain relations, producing the appropriate output.

WHERE ARE WE HEADED?

The restaurant gazetteer example and its generalization and implementation into a mediator plan are just samples of how current extraction and integration tools can be used to automate the discovery, extraction, and integration of geographic-data sources for the production of gazetteers. What this means for the greater research community is that vast quantities of information describing the world, geographically speaking, are at their fingertips, ready to be harvested and exploited for any number and variety of research activities. By iteratively running these newly emerging gazetteer generation tools, intelligence analysts and decision makers can use gazetteers in conjunction with other reasoning systems and tools to trace the movements of people, vehicles, goods, and services.

The automated frameworks being created for these tasks generate immediate results, extracting information from the available data sources, at the instant they are queried, which reflects the view of the world of the source. Storing the previous results of iterative runs enables a full history to be compiled of an area and how it has changed over time, thereby providing valuable information for those interested in the changing character of the geographic domain. For instance, if the workflow presented in Figure 4.1 (the simple restaurant gazetteer example) was scheduled to run every day, the results could be used to track the arrival, departure, and movement of restaurants within an area of interest. This type of information is highly descriptive of the dynamic urban commercial landscape and would be useful for those interested in data describing dynamic processes, such as economic growth patterns (as more restaurants open or close), population movements (as different classes of restaurants catering to different ethnic communities come and go), and land use (as built structures transition from one use class to another).

With a view to increasing the resolution and accuracy of the gazetteer even further, the "Wiki" concept has recently been applied to the gazetteer domain. Several online sites (e.g., http://geonames.org; GeoNames.org 2007) help individuals knowledgeable about a particular geographic area to share their knowledge in a flexible and unofficial manner by simply adding, deleting, and correcting feature entries through an Internet form. The people entering information about local areas (one would hope) are the people who actually live in and are knowledgeable about that area. This method of gazetteer-feature creation and refinement adds an entirely new level of local-scale accuracy and an entirely new set of potential problems. There are (currently) no restrictions about who can enter information about features, so, as famously shown through scandals involving the validity of the data on the original Wikipedia.com site (Seigenthaler 2005), the trust that one should place in the information about geographic features contained in a Wiki gazetteer is questionable. Although the richness of local knowledge is a plus, the unedited nature of the information must be weighed against the official and authorized information available elsewhere. In addition, when the general public is allowed to create gazetteer entries at will, the result will be many different views of a specific place at a specific time. These multiple views will generate many issues that must be addressed prior to their usage in applications that have particular accuracy or reliability constraints. These include topics such as the certainty (accuracy) and vagueness (resolution) of the descriptive attributes as well as confidence in the source.

Possible directions for addressing these and the multitude of other issues created through new gazetteer-generation techniques include applying data-reliability checks such as secondary source validation (Michalowski et al. 2005), quorum voting (Malkhi et al. 1997), and the inclusion of authoritative trusted sources (Gertz et al. 2004). However large the difficulties introduced, it is easy to imagine that the validity of these types of resources might improve over time, and this outcome makes them valuable for their accuracy and resolution as long as the data entered can be assumed to be legitimate.

The preceding subsections have attempted to describe the concept and structure of the gazetteer both in terms of its fundamental content and uses, and in terms of how the emerging data sources and methods leading to its creation are driving it in new, previously unimaginable directions. One key geospatial tool that is fundamentally linked to the improvements in gazetteer completeness and accuracy is the geocoder. In many instances, the geocoder is the tool used to create footprints for geographic features in the gazetteer (as in our prior gazetteer-generation examples), and the strengths and weaknesses of the geocoder used will have numerous effects on the resulting gazetteer (Goldberg et al. 2007). The next section offers an in-depth exploration of this topic.

GROUNDING THE CHANGING WORLD — THE GEOCODER

Today, most gazetteers contain single-point representations for their footprints. Topological relations between features are limited; for example, overlap and containment cannot be derived from points. Most gazetteers include explicit statements of administrative hierarchies (e.g., a city is part of a county is part of a state is part of a country), but these same relationships cannot be derived from the simple point footprints. Large vector features, such as rivers, are often represented by a single point at their mouths, an enormous generalization that greatly reduces the usefulness of the footprints. When large area features such as major cities or countries are represented as single points, the representation is useless for many applications. These shortcomings have recently begun to be addressed through more sophisticated geocoding tools. As we will see, the improvements made to the underlying methodology, the inclusion of new types of data sources, and the expansion (or perhaps better termed, relaxation) of the overarching concept of geocoding all play a role in improving the usefulness and accuracy of geocoding processes.

CREATING A TRUER REPRESENTATION OF THE DYNAMIC WORLD

Improvements made to geocoder tools can most easily be seen through both the accuracy and representation of their outputs. When considering (for simplicity) only point output, more accurate means that the location of the point produced will be closer to where it should be on the ground. Improving the output of geocoders requires rethinking what its output should be, in order to meet user requirements. Although single points are sufficient for some purposes (e.g., orientation of a map view) other analyses require footprints that bound an area or trace a linear feature. Some projects require three-dimensional (3D — length, width, height) or four-dimensional (4D — length, width, height, time) representations. In the case of geocoding an address, this could correspond to returning the 3D structure of the building as it was at a particular time (making it 4D), or the 3D structure of the hills and valleys at a particular time (4D). Other considerations for improving geocoder output are using more accurate reference sources (Ratcliffe 2001, Cayo and Talbot 2003) and eliminating assumptions in the geocoding process (Bakshi et al. 2004, Goldberg et al. 2007).

Many of the same advancements leading to the production of more complete and accurate gazetteers (e.g., availability of data sources, employment of mediator technologies) are driving advances in geocoding technologies on both fronts: accuracy and representation (e.g., Bakshi et al. 2004). This, in turn, is leading to the production of gazetteers whose features are more representative of the real world, with topological relations between features intact. Attacking the very assumptions that have traditionally caused the spatial inaccuracies in the geocoding process, new tools supporting new algorithms for deriving a geographic point from a textual piece of information are vastly improving the results of the geocoding process. Both the address range (Ratcliffe 2001) and parcel homogeneity (Dearwent et al. 2001) assumptions that plague linear interpolation-based geocoding solutions can be avoided by using the same information extraction tools mentioned earlier. Figures 4.6 and 4.7 show how the *GetFootprints()* domain rule can be expanded with new data sources and operations resulting in both highly accurate point representations and alternative polygon footprints depending on user needs. Figure 4.6 conceptually depicts multiple versions of the same function. Figure 4.7 shows how these could be implemented in the mediator system.

By wrapping nontraditional, highly accurate, local-scale information sources that can validate the existence and location of addresses along a street segment (such as local tax assessor Web sites), geocoding tools can generate geographic coordinates for addresses based on the actual number of parcels on the street, rather than by using the less-accurate address ranges associated with street vector sources. Sometimes, the dimensions of parcel lots along a street are available and, with this knowledge in hand, geocoding algorithms no longer need to assume that all parcels along a street segment are the same size (the parcel homogeneity assumption). Instead, the

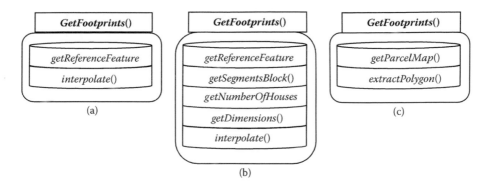

FIGURE 4.6 Alternative expansions of GetFootprints().

algorithm can reason over the length of the street segments and the sizes of the parcels on the entire block (all four sides in the case of a normal square block) to determine the most probable layout for the parcels, including which parcels are on which corners (the so-called corner problem). From this, interpolation can occur within a smaller domain of error, ultimately leading to far greater accuracy for the resulting geocoded location (Bakshi et al. 2004). In the second depiction (Figure 4.7b), this situation is presented. In the third depiction (Figure 4.7c), we see that we can just as

```
GetFootprints($address, $boundingBox:'US', $minAccuracy:'0', feature):–
GetFootprintsLinear ($ address, feature) V
GetFootprintsLinearAdvanced ($ address, feature) V
GetFootprintsExtract ($ address, feature) V
```

```
GetFootprintsLinear ($ address, $boundingBox, feature):–
GetReferenceFeature($address, ref)^
Interpolate($address, $ref, feature)^
boundingBox='US'^
minAccuracy='0'
```

```
GetFootprintsLinearAdvanved ($ address, $boundingBox, feature):–
GetReferenceFeature($address, ref)^
GetSegmentsOnBlock($ref, block)^
GetNumberOfHouses($ref, num)^
GetDimensions($ref, dim)^
Interpolate($address, $ref, $block, $num, $dim, feature)^
boundingBox='LA County'^
minAccuracy='5'
```

```
GetFootprintsExtract($ address, $boundingBox, feature):–
GetParcelMap($address, map)^
ExtractPolygon(map, feature)^
boundingBox='Santa Monica'^
minAccuracy='9'
```

FIGURE 4.7 Expanded GetFootprints() domain models.

```
Tiger($address, ref):–
GetRerferenceFeature($ address)^
GetSegmentsOnBlock($ref, block)^
boundingBox = 'US'
```

```
InterpolationEngine($ address, $ref, feature):–
Interpolate($address, $ref, feature)
```

```
Assessor($ address, $ref, num, dim):–
GetNumberOfHouses($ref, num)^
GetDimensions($ref, dim)^
boundingBox = 'LA County'
```

```
CityMap($ address, map):–
GetParcelMap($address, map)^
boundingBox='Santa Monica'
```

```
ExtractionEngine($ map, feature):–
ExtractPolygon($map, feature)
```

FIGURE 4.8 Expanded GetFootprints() source descriptions.

easily substitute a different composite function that simply extracts the parcel polygon from an online raster map.

The additional (optional) parameters of $boundingBox and $minAccuracy have also been added in Figure 4.7, constraining the execution of each function prototype (i.e., the scope constraints), with default values assumed if absent. The $minAccuracy parameter relates to a relative level of accuracy associated with each geocoding method, 0 being the least accurate and 9 being the most. The $boundingBox parameter relates to the source descriptions defined in Figure 4.8 and defines the area for which each source is applicable. For clarity, textual strings (i.e., "LA County") are used in place of the actual geographic footprints that would be used in practice. When a geocoding query is presented to the geospatial mediator, the system will use the domain rules in Figure 4.7 and the source relations in Figure 4.8 to reason which sources and methods are appropriate to query and execute based on the required level of accuracy and availability of data. In this way, a geospatial mediator-based geocoding system is easily expandable as new data sources become available and new geocoding techniques are developed. The system will automatically determine the best methods and data sources to use based on the attributes associated with each. Figure 4.9 depicts extracting the parcel boundaries from a raster Web-map assessor Web site to return a polygon as the result of the most accurate geocoding method.

It is equally important to note that these additional data sources can be queried every time a geocoding operation is to be performed. Being national-scale and released every 10 years, the Census's TIGER Line files (a commonly used and freely available street-vector data source), like many other data sources used as a reference file for geocoding, represent static views of the world. They are temporally fixed to the time period when they were created, with deteriorating quality as time passes. This unavoidable degradation over time ultimately leads to geocoding results that

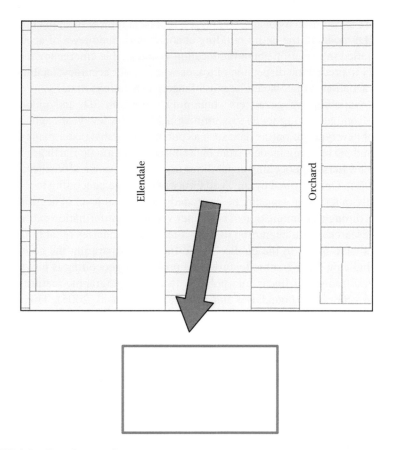

FIGURE 4.9   Parcel extraction.

become more and more inaccurate (Bonner et al. 2003), a fact that is seldom reported along with the data produced (Rushton et al. 2006). In contrast, the newly emerging geocoding tools using multiple sources for feature creation and validation (such as the example in Figures 4.6 and 4.7 based on geospatial mediator technologies) represent the current state of the same geocoded location.

BEYOND TRADITIONAL GEOCODING

As mentioned in the previous sections, different applications will require varying levels of spatial representations and resolutions for the geographic data contained in the gazetteers that they use for their data sources. For instance, in the aforementioned police-blotter example, a simple point representation may be sufficient for the spatial analysis carried out by police investigators into patterns of crime occurrence for a particular area (Chainey and Ratcliffe 2005). In contrast, if the application were geared to more of an emergency-management situation and the police officers responding to a report needed to know information about potential hazardous materials being released into the environment for evacuation purposes, a single-point representation of a building would not suffice. In this case, the information required would be the two-dimensional (2D) polygon representation or even the three-dimensional (3D) model of the entire building at some predefined level of accuracy. This more detailed information about the location would enable emergency management personnel to predict the dispersion of toxins with greater accuracy, enabling them to direct evacuations of people in the most dangerous locations first.

The availability of gazetteers containing subparcel, 3D, and indoor features (derived from geocoding tools) was unimaginable just a few years ago. The proliferation of new types of online data such as publicly available raster maps of assessor parcels and 3D building models from data sources like Google Earth are facilitating entirely new tiers of geocoding capabilities (e.g., Hutchinson and Veenendall 2005a, b; Lee 2004). Once limited by the availability of data, the geocoding tools are growing rapidly in terms of what can be geocoded and to what level of accuracy. As one example, through a combination of computer vision and information extraction techniques, researchers are beginning to extract feature boundaries reliably from online maps available freely to the public via the Internet. Separating the composite layers into individual geographic features, the process of geocoding is moving beyond simple and traditional linear interpolation to actual geographic-feature extraction with true geographic boundaries intact (e.g., Chiang et al. 2005). This represents a significant advance in both the accuracy and representation of the geocoded features produced, providing an unprecedented feature-by-feature view of the changing world. As features change over time, each of these representations can be captured, providing accurate data from which specific events and other complex space-time relationships can be derived.

Additionally, the types of information that can be geocoded are rapidly increasing. Long the bane of geocoding practitioners, for example, sub-parcel geocoding is quickly becoming a reality (Hutchinson and Veenendall 2005a, b; Lee 2004). The importance of this to researchers of all stripes cannot be understated. In the health-research community alone, accurate representations of a person's domicile, down to the apartment level, will serve to promote entire new lines of research that were heretofore unavailable due to the availability of geocoded data at too-low resolution (e.g., down to "somewhere" in the parcel of the apartment complex's footprint). Environmental exposure models sensitive down to meters or tens of meters can now be reliably created, applied, and validated with greater confidence because of the more accurate data being produced (e.g., Ward et al. 2005).

Perhaps even more important, the geocoding of relative spatial locations is proving feasible. Locational descriptions such as "half a mile north of Tutor Hall," being both sub-parcel (Tutor Hall on USC Campus) and relative (half mile), have never been reliably geocoded by any commercial geocoding method. Prototypical research platforms have investigated the possibility, but production systems are highly specialized for a particular area, and operational systems can be found for only a handful of places. Combining fundamental research from the fields of machine learning, artificial intelligence, and GIS, emerging research is providing the framework for this type of tool to become a reality (cf. Hutchinson and Veenendall 2005a, b). When it comes to fruition, this line of research will provide immeasurable utility to the most basic of geographical concepts, explaining the location of something in simple terms. When a call comes into a 911 center, unless a person is at his or her own home or work, he or she will probably not be able to give an exact address where he or she needs assistance, much less the exact geographical location. People speak most easily in terms of relative locations and as such, a tool to geocode them quickly and accurately is of utmost importance. Responding to unexpected events and building credible intelligence reports in a dynamically changing world will necessitate it.

DYNAMIC INTEGRATION

The dynamic modeling of the environment that newly emerging gazetteer and geocoding services provide is just one outcome that these tools afford. In addition to being useful for data generation and representation, these tools are also facilitating the integration of disparate, dynamic data sources for use in the analysis and interpretation of the changing world they represent (cf. Shahabi et al. 2006). Each of these tools brings different strengths to the problem of dynamic data integration and they therefore complement one another.

ENABLING INTEGRATION WITH GAZETTEERS

The primary strength of the classical gazetteer conceptual structure is that it enables the identification of a geographic feature from information along any of its three axes (name, type, or footprint) (Hill 2000). This structure enables the gazetteer to assume a central role in dynamic data integration. First and foremost, the gazetteer possesses the ability to link a variety of different geographic references due to the simple fact that any information about a geographic feature of interest must refer to at least one of the three axes. For example, the gazetteer could be used to link dynamically changing data about political stability with the probable locations of one or more political activists. Alternatively, a gazetteer might be used to identify all of the stream gauging stations based on the feature types in a hydrological data source that are downstream of a dam that is likely to fail.

We can see through the previous examples that, because the gazetteer cuts across the full spectrum of possible information that can be used for describing geographic features (e.g., spatially, temporally, toponymically, typographically), it can be used as the basis for reasoning with a wide variety of geospatial data sources. By integrating a data set with the gazetteer (e.g., intersecting along the appropriate axis of the gazetteer), it becomes immediately and automatically grounded to the other axes as well. A set of place names can be associated with spatial footprints and types simply by intersecting along the toponymic axis. Likewise, a set of spatial footprints can have names and types associated with them by integrating the two sets of geometries. While there may be cases where the automatic association of information might be to the wrong scale or resolution to be useful (e.g., every parcel in a city will have the name "United States" associated with it if intersected spatially with a gazetteer of world countries), this same fundamental ability can prove invaluable when applied to the correct situation using the correct data sources.

Likewise, advancements made to geocoding services are enabling immediate access to unprecedented amounts of information based on address data. An often-cited quotation used primarily to validate the time and money being spent on geocoding research projects is that "80–90% of [U.S.] government data is geographic data" (Federal Geographic Data Committee 2006), with the majority of that consisting of raw address data or geocoded address data (Croner 2003). From this standpoint alone, we could rightfully conclude that address data could be considered among the most (if not the most) ubiquitous form of geospatial data in existence. Like never before, geographic references that lead to any form of addressable data can now be integrated immediately with other spatial data sets, through the contents of a gazetteer database and/or the use of geocoding technologies to convert a textual address (nongeographic data) into some form of geospatial data (e.g., a point, line, or feature).

CONCLUSIONS

Taken together, the newly emerging gazetteer and geocoding tools being developed provide a powerful combination with which to both describe and work with dynamic data in a changing world. Fundamental research advances in both of these interrelated fields are leading the way in both information generation and information integration, each of which are of particular concern to intelligence analysts and decision makers as they try to explain the trends and events that characterize our continually evolving world.

REFERENCES

Adali, S., K. S. Candan, Y. Papakonstantinou, and V. S. Subrahmanian. 1996. VS. Query caching and optimization in distributed mediator systems. In *Proceedings of the 1996 ACM SIGMOD International Conference on Management of Data*, 137–46. New York: ACM Press.

Agarwal, P. 2004. Contested Nature of Place: Knowledge Mapping for Resolving Ontological Distinctions Between Geographical Concepts. In *GIScience 2004*, ed. M. J. Egenhofer, C. Freksa, and H. J. Miller, 1–21. Lecture Notes in Computer Science 3234. Berlin: Springer-Verlag.

Agouris, P., K. Beard, G. Mountrakis, and A. Stefanidis. 2000. Capturing and Modeling Geographic Object Change: A SpatioTemporal Gazetteer Framework. *Photogrammetric Engineering and Remote Sensing* 66(10):1224–50.

Alexandria Digital Library. 2007a. http://www.alexandria.ucsb.edu/. Last accessed 3/25/2007.

Alexandria Digital Library. 2007b. *Guide to the ADL Gazetteer Content Standard*, Version 3.2. http://www.alexandria.ucsb.edu/gazetteer/ContentStandard/version3.2/GCS3.2-guide.htm. Last accessed 3/25/2007.

Ambite, J. L. and C. A. Knoblock. 1997. Planning by Rewriting: Efficiently Generating High-Quality Plans. In *Proceedings of the 14th National Conference on Artificial Intelligence*. Providence, RI: AAAI Press / MIT Press. 706–13.

Axelrod, A. 2003. On Building a High Performance Gazetteer Database. In A. Kornai and B. Sundheim (eds.): HLT/NAACL '03: *Proceedings of Workshop on the Analysis of Geographic References held at Joint Conference for Human Language Technology and Annual Meeting of the North American Chapter of the Association for Computational Linguistics*. Boston, MA: Association for Computational Linguistics. 63–68.

Bakshi, R., C. A. Knoblock, and S. Thakkar. 2004. Exploiting Online Sources to Accurately Geocode Addresses. In D. Pfoser, I.F. Cruz, and M. Ronthaler (eds.): *ACM-GIS '04: Proceedings of the 12th ACM International Symposium on Advances in Geographic Information Systems.* Washington, D.C.: ACM Press. 194–203.

Barish, G. and C. A. Knoblock. 2005. An Expressive Language and Efficient Execution System for Software Agents. *Journal of Artificial Intelligence Research* 23:625–66.

Bonner, M. R., D. Han, J. Nie, P. Rogerson, J. E. Vena, and J. L. Freudenheim. 2003. Positional Accuracy of Geocoded Addresses in Epidemiologic Research. *Epidemiology* 14(4). 408–11.

Buckland, M. and L. Lancaster. 2004. Combining Place, Time, and Topic: The Electronic Cultural Atlas Initiative. *D-Lib Magazine* 10(5). http://www.dlib.org/dlib/may04/buckland/05buckland.html.

Cayo, M. R. and T. O. Talbot. 2003. Positional error in automated geocoding of residential addresses. *International Journal of Health Geographics* 2(10).

Chainey, S. and J. Ratcliffe. 2005. *GIS and Crime Mapping.* Chichester, UK: Wiley & Sons.

Chiang, Y-Y., C. A. Knoblock, and C-C. Chen. 2005. Automatic extraction of road intersections from raster maps. In *The 13th ACM International Symposium on Advances in Geographic Information Systems* (ACM-GIS'05), Bremen, Germany, November. 267–76.

Croner, C. M. 2003. Public Health GIS and the Internet. *Annual Review of Public Health* 24: 57–82.

Dearwent, S. M., R. R. Jacobs, and J. B. Halbert. 2001. Locational uncertainty in georeferencing public health datasets. *Journal of Exposure Analysis Environmental Epidemiology* 11(4):329–34.

Doerr, M. 2001. Semantic Problems of Thesaurus Mapping. *Journal of Digital Information* 1(8). 2001–2003.

Doerr, M. and M. Papagelis. 2006. A method for estimating the precision of place name matching. Unpublished, under review to *International Journal of Digital Libraries.* http://www.ics.forth.gr/isl/publications/paperlink/ecdl.pdf. Last accessed 11/24/2006.

Dugandzic, R., L. Dodds, D. Stieb, and M. Smith-Doiron. 2006. The association between low level exposures to ambient air pollution and term low birth weight: a retrospective cohort study. *Environmental Health* 5 (3).

Federal Geograhpic Data Committee. 2006. Homeland Security and Geographic Information Systems: How GIS and mapping technology can save lives and protect property in post-September 11th America. http://www.fgdc.gov/library/whitepapers-reports/whitepapers/homeland-security-gis. Last accessed 11/24/2006.

Frew, J., M. Freeston, N. Freitas, L. L. Hill, G. Janée, K. Lovette, R. Nideffer, T. Smith, and Q. Zheng. 1998. The Alexandria Digital Library Architecture. Research and Advanced Technology for Digital Libraries. In *Proceedings of the Second European Conference* (ECDL). 61–73.

Frew, J., M. Freeston, N. Freitas, L. L. Hill, G. Janée, K. Lovette, R. Nideffer, T. Smith, and Q. Zheng. 2000. The Alexandria Digital Library Architecture. *International Journal on Digital Libraries* 2(4):259–68.

Frank, A. U. 2001. Tiers of ontology and consistency constraints in geographical information systems. *International Journal of Geographical Information Science* 15(7):667–78.

Gadia, S. K. 1998. A homogeneous relational model and query languages for temporal databases. *ACM Transactions on Database Systems* 13(4). New York, NY: ACM Press. 418–48.

GeoNames.org. 2007. http://www.geonames.org, Last accessed 3/25/2007.

Gertz, M., M. T. Ozsu, G. Saake, and K. U. Sattler. 2004. Report on the Dagstuhl Seminar: Data quality on the Web. *SIGMOD Record* 33(1). New York, NY: ACM Press. 127–32.

Goldberg, D. G., J. P. Wilson, and C. A. Knoblock. 2007. From Text to Geographic Coordinates: The Current State of Geocoding. *URISA Journal* 19(1):33–46.

Goodchild, M. 1999. The future of the gazetteer. Unpublished, presented at the Digital Gazetteer Information Exchange Workshop. http://www.alexandria.ucsb.edu/~lhill/dgie/DGIE_website/DGIE final report.htm. Last accessed 11/24/2006.

Goodchild, M. and L. L. Hill. 2007. *Summary Report: Digital Gazetteer Research & Practice Workshop*, December 7–9, 2006, Santa Barbara, CA. http://www.ncgia.ucsb.edu/projects/nga/docs/DGRP-summary-report.pdf. Last Accessed 3/25/2007.

Gupta, A., R. Marciano, I. Zaslavsky, and C. Baru. 1999. Integrating GIS and Imagery through XML-Based Information Mediation. *Integrated Spatial Databases: Digital Images and GIS*. Volume 1737 of Lecture Notes in Computer Science. Berlin: Springer. 211–34.

Hill, L. L., J. Frew, and Q. Zheng. 1999. *Geographic names: The implementation of a gazetteer in a georeferenced digital library*. Technical report.

Hill, L. L. and Zheng Q. 1999. Indirect geospatial referencing through place names in the digital library: Alexandria digital library experience with developing and implementing gazetteers. In *62nd Annual Meeting of the American Society for Information Science*. Medford, NJ: Information Today. 57–69.

Hill, L. L. 2000. Core elements of digital gazetteers: Placenames, categories, and footprints. In J. L. Borbinha and T. Baker (eds.): *ECDL '00: Research and Advanced Technology for Digital Libraries, 4th European Conference*. Volume 1923 of Lecture Notes in Computer Science. London, UK: Springer. 280–90.

Hill, L. L. 2006. *Georeferencing: The Geographic Associations of Information*. Cambridge, MA: MIT Press.

Himmelstein, M. 2005. Local Search: The Internet Is the Yellow Pages. *Computer* 38(2). Piscataway, NJ: IEEE Press. 26–34.

Hornsby, K. 2001. Temporal Zooming. *Transactions in GIS* 5(3):255–72.

Hutchinson, M. and B. Veenendall. 2005a. Towards a Framework for Intelligent Geocoding. In *SSC 2005 Spatial Intelligence, Innovation and Praxis: The National Biennial Conference of the Spatial Sciences Institute*.

Hutchinson, M. and B. Veenendall. 2005b. Towards Using Intelligence to Move from Geocoding to Geolocating. In *7th Annual GIS in Addressing Conference*. Park Ridge, IL: Urban and Regional Information Systems Association.

Knoblock, C.A., K. Lerman, S. Minton, and I. Muslea. 2001. A machine-learning approach to accurately and reliably extracting data from the Web. In *Proceedings of the IJCAI Workshop on Adaptive Text Extraction and Mining*.

Laender, A.H.F., K.A.V. Borges, J.C.P. Carvalho, C. B. Medeiros, A. S. da Silva, and C. A. Davis Jr. 2005. In Y. Manalopoulos and A. N. Papadapoulos (eds.): Integrating Web Data and Geographic Knowledge into Spatial Databases. *Spatial Databases: Technologies, Techniques and Trends*. Chapter 2:23–47.

Laender, A.H.F., B. Ribeiro-Neto, A. Silva, and J. Teixeira. 2002. A Brief Survey of Web Data Extraction Tools. *SIGMOD Record* 31(2). New York, NY: Association for Computing Machinery. 84–93.

Lam, N. S-N. and D. A. Quattrochi. 1992. On the issues of scale, resolution, and fractal analysis in the mapping sciences. *The Professional Geographer* 44(1):88–98.

Lee, J. 2004. 3D GIS for Geo-coding Human Activity in Micro-scale Urban Environments. In M. J. Egenhofer, C. Freksa, and H. J. Miller (eds.): Geographic Information Science: Third International Conference, *GIScience 2004*, 162–78. Berlin: Springer-Verlag.

Malkhi, D., M. Reiter, and R. Wright. 1997. Probabilistic quorum systems. In *PODC '97: Proceedings of the sixteenth annual ACM symposium on principles of distributed computing*. New York, NY: ACM Press. 267–73.

Maynard, D., K. Bontcheva, and H. Cunningham. 2004. Automatic Language-Independent Induction of Gazetteer Lists. In *Proceedings of 4th Language Resources and Evaluation Conference* (LREC'04).

Michalowski, M., S. Thakkar, and C. A. Knoblock. 2005. Automatically utilizing secondary sources to align information across sources. *AI Magazine* 26(1). Menlo Park, CA: AAAI Press. 33–44.

Mikheev, A., M. Moens, and C. Grover. 1999. Named entity recognition without gazetteers. In H. S. Thompson and A. Lascarides (eds.): *EACL '99: 9th Conference of the European Chapter of the Association for Computational Linguistics*. San Francisco, CA: Morgan Kaufmann Publishers. 1–8.

Naumann, F. and M. Haeussler. 2002. Declarative Data Merging with Conflict Resolution. In *Proceedings of the International Conference on Information Quality* (IQ 2002).

Plewe, B. 2002. The Nature of Uncertainty in Historical Geographic Information. *Transactions in GIS* 6(4):431–56.

Raman, V. and J. M. Hellerstein. 2002. Partial results for online query processing. In SIGMOD '02: *Proceedings of the 2002 ACM SIGMOD international conference on management of data*. New York, NY: ACM Press. 275–86.

Ratcliffe, J. H. 2001. On the accuracy of TIGER-type geocoded address data in relation to cadastral and census areal units. *International Journal of Geographical Information Science* 15(1):473–85.

Rushton, G., M. Armstrong, J. Gittler, B. Greene, C. Pavlik, M. West, and D. Zimmerman. 2006. Geocoding in Cancer Research — A Review. *American Journal of Preventive Medicine* 30(2):S16–S24.

Seigenthaler, J. 2005. A false Wikipedia 'biography.' USA Today. http://www.usatoday. com/news/opinion/editorials/2005-11-29-wikipedia-edit_x.htm.

Shahabi, C., Y-Y. Chiang, K. Chung, K-C. Huang, J. Khoshgozaran-Haghighi, C. A. Knoblock, S. C. Lee, U. Neumann, R. Nevatia, A. Rihan, S. Thakkar, and S. You. 2006. GeoDec: Enabling Geospatial Decision Making. In *Proceedings of the IEEE International Conference on Multimedia & Expo* (ICME).

Shimada, S. and H. Fukui. 1999. Geospatial Mediator Functions and Container-Based Fast Transfer Interface in Si3CO Test-Bed. In *Proceedings of the Second International Conference on Interoperating Geographic Information Systems*. London, UK: Springer-Verlag. 265–76.

Smith, D.A. and G. S. Mann. 2003. Bootstrapping toponym classifiers. In A. Kornai and B. Sundheim (eds.): *HLT/NAACL '03: Proceedings of Workshop on the Analysis of Geographic References held at Joint Conference for Human Language Technology and Annual Meeting of the North American Chapter of the Association for Computational Linguistics*. Boston, MA: Association for Computational Linguistics.

Thakkar, S., J. L. Ambite, and C. A. Knoblock. 2005. Composing, optimizing, and executing plans for bioinformatics web services. *VLDB Journal* 14(3):330–53.

Turmo, J., A. Ageno, and N. Català. 2006. Adaptive information extraction. *ACM Computing Surveys* 38(4). New York, NY: ACM Press.

U.S. Board on Geographic Names. 2006. Geographic Names Information System. http://geonames.usgs.gov/pls/gnispublic. Last accessed 1/24/2006.

U.S. Central Intelligence Agency. 2007. *The World Factbook, 2007*. Washington, D.C.: U.S. Central Intelligence Agency.

U.S. National Geospatial-Intelligence Agency. 2006. GeoNet Names Server. http://gnswww. nga.mil/geonames/GNS. Last accessed 11/24/2006.

Van Rijsbergen, C. J. 1979. *Information Retrieval*. Newton, MA, USA: Butterworth-Heinemann.

Wang, J. and N. Ge. 2006. Automatic feature thesaurus enrichment: extracting generic terms from digital gazetteer. In *JCDL '06: Proceedings of the 6th ACM/IEEE-CS joint conference on Digital libraries*. New York, NY: ACM Press. 326–33.

Ward, M. H., J. R. Nuckols, J. Giglierano, M. R. Bonner, C. Wolter, M. Airola, W. Mix, J. S. Colt, and P. Hartge. 2005. Positional Accuracy of Two Methods of Geocoding. *Epidemiology* 16(4):542–47.
WikMapia.org. 2007. http://www.wikimapia.org, Last accessed 3/25/2007.
Yahoo! Inc. 2006. Yahoo! Local Pages. http://local.yahoo.com. Last accessed 11/24/2006.

5 Reconstructing Individual-Level Exposure to Environmental Contaminants Using Time-GIS

Jaymie R. Meliker

CONTENTS

"I sometimes ask myself how it came about that I was the one to develop the theory of relativity. The reason, I think, is that a normal adult never stops to think about problems of space and time. These are things which he has thought about as a child. But my intellectual development was retarded, as a result of which I began to wonder about space and time only when I had already grown up."

— **Albert Einstein**

INTRODUCTION

The past two decades have witnessed a dramatic increase in the availability of digi-
tally georeferenced data and the capability of Geographic Information Systems (GIS)
to display those data. The importance of GIS for medical research and epidemiology
has long been recognized (Barnes and Peck 1994; Clarke et al. 1996; Croner et al.
1996), with John Snow's map of cholera and drinking water in mid-19th-century
London perhaps the most famous example (Baker and Taras 1981). In recent years,
GIS have increasingly been used for reconstructing individual-level exposure to environmental contaminants in epidemiologic research (Beyea and Hatch 1999; Nuckols
et al. 2004; Ward et al. 2000). The growing body of literature on this application
most commonly includes studies that assess proximity of individuals to sources of
environmental contaminants such as pesticide application on farms (Reynolds et al.
2005), landfill sites (O'Leary et al. 2004), and hazardous waste sites (Elliott et al.
2001; McNamee and Dolk 2001). On occasion, investigators implement environmental fate and transport models to improve the exposure assessment (Reif et al.
2003). These studies, however, rely almost exclusively on home residence at time of
interview or diagnosis as the spatial location of the individual. Yet individuals are
mobile and frequently change residences, and this residential mobility information is
crucial for spatial analyses of chronic diseases with relatively long induction periods,
such as cancer.

The time-geographic concept of mobility histories has been around for at least
several decades (Hagerstand 1970) and these trajectories are sometimes referred to as
geospatial lifelines (Hornsby and Egenhofer 2002; Mark and Egenhofer 1998; Mark
et al. 2000; Sinha and Mark 2005). These lifelines, however, are seldom included in
GIS-based exposure assessments, a consequence, at least in part, of the atemporal
nature of GIS. GIS operate within a static worldview that is largely incapable of
representing temporal change (Goodchild 2000). GIS are best suited to "snapshots"
of static systems (Hornsby and Egenhofer 2000) that hinder the mapping, representation, and analysis of dynamic health, socioeconomic, and environmental information
for mobile populations. In the few GIS-based exposure assessments where residential
histories are included, researchers create numerous maps, or snapshots, in specified
time intervals (usually annual) to assess exposure as individuals move through time
(Aschengrau et al. 1996; Bellander et al. 2001; Bonner et al. 2005; Brody et al. 2002;
Nyberg et al. 2000; Stellman et al. 2003; Swartz et al. 2003). Assembling these snapshots, however, is plagued by a host of problems: (a) it is dataset-intensive, requiring
a unique database and map for each time slice; (b) it is labor-intensive and as such
has the potential to produce critical errors during data manipulation; (c) information
about change is not available in the interval between two consecutive snapshots;
(d) it is so time-consuming that it typically only enables one attempt at exposure
reconstruction — it does not allow for improvements in the underlying models of
environmental contamination for refining and recalculating exposure in an iterative
manner; and (e) dozens-to-hundreds of maps need to be created to calculate time-
resolved exposure using different temporal orientations, such as participants' age,
calendar year, and years prior to diagnosis/interview (see section on temporal orientations for further discussion). As a result of these challenges and limitations in GIS,

most researchers conducting GIS-based exposure assessments choose to disregard residential histories, thereby implicitly assuming that individuals are immobile, and that the induction period between causative exposures and health events (e.g., diagnosis, death) is negligible (Jacquez 2004). The availability of software that efficiently and easily displays and analyzes smooth, continuous space-time datasets, therefore, could profoundly improve individual lifetime-exposure reconstruction (AvRuskin et al. 2004; Meliker et al. 2005).

I am not alone in identifying these weaknesses in GIS-based exposure assessments. In a recent review of epidemiological methods for evaluating geographic exposures and hazards, Mather et al. (2004) lamented the scarcity of methods that account for residential histories of cases and controls. At a meeting of this nation's experts on the spatial analysis of cancer data, the need to account for latency and human mobility in cancer studies was recognized as the second-most pressing issue (Pickle et al. 2005). As Pickle and colleagues (2006) summarize from that meeting, " . . . few tools and methods can be applied to both space and time together . . . a data representation problem needs to be solved: how to store and retrieve integrated space-time data consisting of multiple sets of data from fundamentally different space-time frames." This problem exists because, despite modern computer technologies for storing and managing temporal and spatial datasets, surprisingly few tools are available for working with space-time datasets (Beaubroef and Breckenridge 2000; Dragicevic and Marceau 2000). Without tools for visualizing and analyzing space-time datasets, participant mobility, changes in contaminant concentrations, and other forms of space-time variability are inadequately incorporated in human exposure assessments.

In response to this need, Time-GIS technology has recently been developed that takes advantage of the space-time variability inherent in many datasets (AvRuskin et al. 2004; Greiling et al. 2005; Jacquez et al. 2005; Meliker et al. 2005; Weaver et al., in press). While traditional GIS are based on spatial data structures — the "what, where" dyad that inadequately displays changes through time, Time-GIS are based on space-time data structures (Galton 2003) that enable characterization of the "what, where, when" triad needed for effective representation of data used to analyze health outcomes. One example of a Time-GIS is STIS™ (Space-Time Intelligence System, TerraSeer, Inc., Crystal Lake, Illinois), which relies on space-time data structures following the concept of temporal intervals adopted by Rodriguez and colleagues (2004) in their analysis of temporal relationships. Using this concept, events, processes, states, and actions are not distinguished; all activities are represented using temporal intervals in the space-time data structure. Time-GIS therefore allow the user to observe and quantify how geographies change with time, enabling powerful exposure reconstruction procedures that are not possible through "space only" GIS.

In this chapter, different facets of Time-GIS functionality will be discussed within the context of individual exposure reconstruction. These include: (1) importing space-time datasets of human mobility patterns and environmental contaminants, (2) integrating information across space-time vector and space-time raster maps, (3) calculating exposure estimates over the life-course, (4) calculating lifetime-exposure estimates using different temporal orientations (e.g., calendar year or participant's age), and (5) validating lifetime exposure and quantifying uncertainty in the exposure estimate. Where possible, Time-GIS functionality will be illustrated using STIS™ software for the example of individual lifetime exposure to inorganic arsenic in southeastern Michigan (Meliker et al. 2007).

IMPORT SPACE-TIME DATASETS OF HUMAN MOBILITY PATTERNS AND ENVIRONMENTAL CONTAMINANTS

One of the space-time datasets necessary for geographic exposure reconstruction is that of human mobility patterns. Human mobility can be conceptualized using multiple representations depending on the time scale under consideration. For example, over the course of a lifetime, mobility patterns are most commonly represented by residential and occupational histories and are often available through self-report or centralized registry (not available in the United States but maintained in several European countries) (Meliker et al. 2007; Storm et al. 1997). These datasets typically include information about changes using the scale of a year, and refer to locations marked by home and work. In contrast, daily-activity patterns typically constitute the sequence and length of time at locations over the course of a day. These types of human mobility datasets are increasingly available thanks to global positioning system (GPS) technology (Elgethun et al. 2003; Phillips et al. 2001). Even though GPS datasets are conceptually thought of as being continuous, output from a GPS is a vector dataset of geographic coordinates associated with discrete temporal intervals (start and end times), similar to how data are stored for residential and occupational histories.

In addition to human mobility, space-time datasets of environmental histories are requisite for geographic exposure reconstruction. Environmental epidemiologists and exposure scientists are increasingly working with raster and vector files of crop type, soil type, water quality, and vector density derived from satellite imagery; results from geostatistical and environmental fate and transport models of air, water, and soil contaminants; and spatially constructed datasets that depict historical environmental changes (e.g., histories of industrial operations, public water supplies, etc.) (Brownstein et al. 2003; Meliker et al. 2007; Reif et al. 2003; Ross et al. 2006; Rull et al. 2006). Maps generated from these efforts typically have multiple time stamps, on a scale ranging from hours to years, and appear in chronological sequence.

As mentioned in the Introduction, visualizing the dynamics inherent in these space-time datasets is not straightforward in a GIS environment, where the user is forced to choose time-slices and create multiple spatial-only maps for data viewing and analysis. One of the difficulties with these spatial-only maps is that information between time-slices is removed and therefore not available for analysis (Weaver et al. in press). Advances in Time-GIS technology, however, have the potential to display space-time datasets in a temporally continuous fashion using an automated import procedure (Table 5.1). Some of this functionality can be demonstrated using vector datasets in the STIS™ software platform. For example, Figures 5.1 and 5.2 display snapshots from temporally continuous space-time maps of residential histories of study participants, and arsenic concentrations in public water supplies (for more detail on the datasets behind these maps and additional discussion of space-time datasets, see Meliker et al. 2005; Meliker et al. 2007). These figures are not simply displaying a series of animated maps, but rather, each figure displays one time-geography. Only one dataset is used to generate each of the animations used in Figures 5.1 and 5.2 (Table 5.2). This is accomplished by using a space-time data structure based on temporal intervals: each row represents a space-time intersection and a variable of interest, and when a geographic location or a variable of interest changes value, a new row is created (Table 5.2). Changes can occur in both attributes and geographies (Peuquet 2001), as is apparent in Figure 5.2, where arsenic concentrations change in some towns' water supplies, along with changes in town boundaries. It is important to re-emphasize that with limitations of the printed page, only static images can be presented here; however, time-GIS produce continuous space-time animations requiring only one dataset for each time-geography.

INTEGRATE INFORMATION ACROSS SPACE-TIME VECTOR AND SPACE-TIME RASTER MAPS

After space-time datasets of human mobility and environmental contamination have been imported into a Time-GIS, a likely next step is to integrate information across these datasets. At least two common procedures handle this functionality in a GIS environment that could translate effectively into a Time-GIS environment: a spatial join method and proximity calculations. In Time-GIS, these functionalities would be applied for every unique space-time representation of the dataset. For example, individuals change locations over the course of a day and are exposed to different concentrations of air pollutants depending on when and where they are located. Given datasets of individual daily mobility patterns and hourly concentrations of nitrogen dioxide in the air, a space-time join function could assign the value of the nitrogen dioxide concentration to each person at every location for the appropriate temporal interval. In another example taking advantage of land-cover classifications (e.g., available in several time-slices from Landsat satellite imagery) and residential histories, the proximity of each individual to farmland could be calculated, again, at every location for the appropriate time interval. Estimates of air-pollution concentrations and proximity to farmland are common measures of exposure in environmental epidemiologic studies (Bellander et al. 2001; Swartz et al. 2003); however, prior to development of Time-GIS technology, temporal dimensions of these geographic exposure assessments were simplified by only relying on time-slices.

An automated space-time join method was demonstrated in a recent lifetime arsenic-exposure assessment calculated using STIS™ (Meliker et al. 2007). In this example, arsenic concentrations were assigned to each individual's home drinking water by successively looping through each person's residential history (Figure 5.1) and assigning an arsenic value for each space-time interval. The estimate of arsenic concentration in home drinking water came from a geostatistically derived raster image for private-well arsenic concentration (Goovaerts et al. 2005), and the public supply space-time map shown in Figure 5.2. Every participant was assigned an arsenic concentration for each residence and for each change in water supply at a residence. This results in an estimate of arsenic concentration in home drinking water

TABLE 5.1

Comparative Functionality of Traditional GIS and Time-GIS for Exposure Reconstruction

Goal	Functionality of Traditional GIS	Functionality of Time-GIS	Implications for Exposure Assessment	Example Application
Import space-time datasets of human mobility patterns and environmental contaminants.	Read-in spatial raster and vector datasets.	Read-in space-time raster and vector datasets.	Datasets of human mobility and concentrations of environmental contaminants change across space and time. No longer need to import only time-slices of these datasets. Entire space-time dataset could be imported.	Import space-time maps of environmental contaminants from geostatistical models and air-pollution dispersion models (e.g., AIRVIRO).
Integrate information across space-time vector and space-time raster maps.	Assign values and proximity metrics based on intersection of spatial datasets.	Assign values and proximity metrics based on intersection of space-time datasets.	Generate more reliable estimates of the intersection of human mobility with changing maps of environmental contaminants. No longer need to rely on snapshots of this intersection, in which time between snapshots is lost.	Allow for human mobility and changing locations of farmland, and calculate proximities of residences to farmland over the life-course.
Calculate exposure over the life-course.	Derive new datasets using spatially varying attributes.	Derive new datasets using spatially, temporally, and spatiotemporally varying attributes.	Calculate estimates of individual lifetime exposure to environmental contaminants.	Multiply spatiotemporally varying estimates of a drinking water contaminant by temporally varying estimates of water consumption to generate exposure estimates.

Calculate lifetime-exposure estimates using different temporal orientations.	Derive new datasets at one snapshot in time, using one temporal orientation.	Derive new datasets incorporating spatiotemporal variability, using multiple temporal orientations.	By relating each individual's age to a calendar year, exposures can be compared among individuals using their ages or calendar years. Similarly, in epidemiologic studies, exposures can be compared using years prior to diagnosis as the temporal orientation.	Exposure estimates over each individual's life-course can be calculated and exported using age, calendar year, or years prior to diagnosis as the temporal orientation. Different temporal orientations can produce distinct results in epidemiologic analyses of the exposure–disease relationship.
Validate lifetime exposure and quantify uncertainty in the exposure estimate.	None available	Derive temporally specific measures of uncertainty and validity (e.g., sensitivity, specificity, predictive value, correlation).	Use uncertainty and validation analyses to compare Time-GIS-based exposure estimate with (1) traditional GIS-based exposure estimate and (2) measurements or biomarkers of individual-level exposure to assess reliability of the Time-GIS-based approach.	For estimates of individual-level air-pollution exposure, compare Time-GIS-based results with (1) traditional GIS-based results and (2) personal air filters to quantify the added benefit from incorporating space-time variability in the exposure assessment.

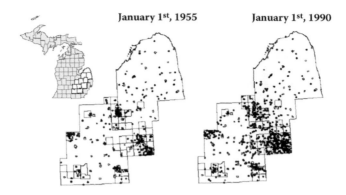

FIGURE 5.1 Residential histories of study participants in southeastern Michigan: snapshots from 1955 and 1990 of continuous animation.

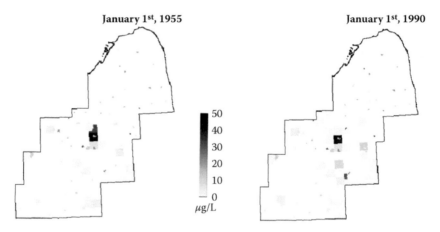

FIGURE 5.2 Arsenic concentrations in public water supplies in southeastern Michigan: snapshots from 1955 and 1990 of continuous animation.

for each moment in a person's life; snapshots of the continuous dataset generated by this space-time join method appear in Figure 5.3.

To attain this same degree of completeness in an exposure assessment using GIS, a new map would need to be constructed for every change in location, or change in attribute at a given location (Table 5.1). This is impractical, inefficient, and error-prone, because there are hundreds to thousands of space-time changes to incorporate into a spatially and temporally resolved exposure assessment in an epidemiologic study. Automating the integration of space-time datasets using Time-GIS enables an exposure scientist to easily and efficiently incorporate space-time variability into an exposure assessment.

TABLE 5.2
Temporal Interval Format for Importing Space-Time Vector Datasets into Time-GIS: Example Using Residential Histories and Water Source History of Participants.

Participant ID	Year Moved In	Year Moved Out	Address*	X-Coord*	Y-Coord*	Water Source	Water Treatment
001239	1/1/1941	1/1/1959	Address #1	694980	264132	Private Well	Water Softener
001239	1/1/1959	1/1/1970	Address #2	687299	268878	Community Supply	Water Softener
001239	1/1/1970	1/1/1994	Address #3	694161	272042	Community Supply	Water Softener
001239	1/1/1994	1/1/2004	Address #4	680421	278791	Private Well	Water Softener
001240	1/1/1946	1/1/1948	Address #1	649645	275342	Community Supply	None
001240	1/1/1948	1/1/1950	Address #2	692980	168978	Community Supply	None
001240	1/1/1950	1/1/2004	Address #3	687699	174042	Community Supply	None
001243	1/1/1953	1/1/1961	Address #1	692161	176791	Community Supply	Water Softener
001243	1/1/1961	1/1/1971	Address #2	660421	177342	Private well	Water Softener
001243	1/1/1971	1/1/1984	Address #3	684656	274665	Private well	Water Softener
001243	1/1/1984	1/1/1993	Address #4	694766	278743	Private well	Water Softener
001243	1/1/1993	1/1/2004	Address #5	686910	274183	Private well	Water Softener + Reverse Osmosis

* Address and geographic coordinates are altered to protect participant confidentiality.

CALCULATE EXPOSURE OVER THE INDIVIDUAL LIFE-COURSE

In addition to intersecting space-time datasets, exposure scientists might wish to combine space-time datasets with temporal datasets to calculate a broad range of individual-level metrics of exposure to environmental contaminants. For example, in the arsenic-exposure assessment calculated using STIS™ (Meliker et al. 2007), arsenic concentration in home drinking water (Figure 5.3) was multiplied by self-reported estimates of water-consumption history to produce individual lifetime

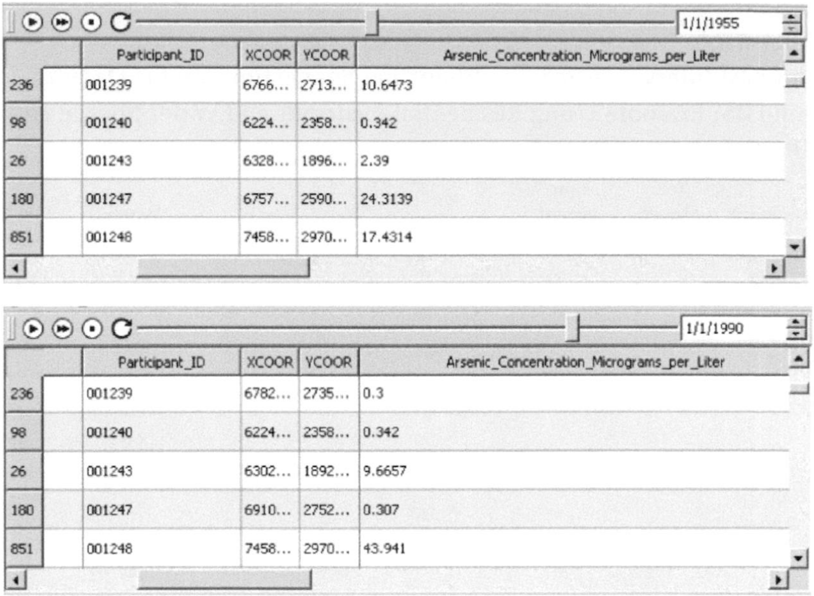

▶ ⏭ ⊙ C	────────────	⎸───────────	1/1/1955 ⇕	
	Participant_ID	XCOOR	YCOOR	Arsenic_Concentration_Micrograms_per_Liter
236	001239	6766...	2713...	10.6473
98	001240	6224...	2358...	0.342
26	001243	6328...	1896...	2.39
180	001247	6757...	2590...	24.3139
851	001248	7458...	2970...	17.4314

▶ ⏭ ⊙ C	────────────	─────────⎸──	1/1/1990 ⇕	
	Participant_ID	XCOOR	YCOOR	Arsenic_Concentration_Micrograms_per_Liter
236	001239	6782...	2735...	0.3
98	001240	6224...	2358...	0.342
26	001243	6302...	1892...	9.6657
180	001247	6910...	2752...	0.307
851	001248	7458...	2970...	43.941

FIGURE 5.3   Arsenic concentrations in individual home drinking water: results of space-time join method. Snapshots from 1955 and 1990 of continuous animation. Snapshots of dataset view in TerraSeer® STIS™. Note: X and Y coordinates are hidden to protect participant confidentiality.

estimates of daily exposure to arsenic in home drinking water (μg/day). The resulting arsenic exposure estimates change over time.

In a general sense, a dataset calculator can derive new variables by applying mathematical functions to both space-time and temporal datasets in a Time-GIS. Individual measures (or estimates) of breathing rates, showering frequency, water consumption, and so on, may be imported into a time-GIS, and combined with human mobility patterns and maps of environmental contaminants to estimate exposure. This can result in the calculation of temporally resolved exposure metrics through inhalation, dermal, and ingestion routes. These exposure estimates will allow environmental epidemiologists and risk assessors to investigate the temporal relationship between exposure and disease in greater detail than is currently possible.

CALCULATE LIFETIME EXPOSURE ESTIMATES USING DIFFERENT TEMPORAL ORIENTATIONS

Another advantage of Time-GIS compared with traditional GIS is data visualization and analysis using multiple temporal orientations. Up to now, the concept of "time" has been used without being explicitly defined. In practice there are several ways that time may be considered when reconstructing exposure: in reference to age (i.e., years/months since being born), in reference to diagnosis/interview (i.e., years/months prior to diagnosis/interview), or in reference to calendar year/date. The temporally specific exposures may differ when one uses these three temporal orientations. For example, if exposures are assessed based on where individuals lived when they were born (age=0), individuals born in different calendar years will be included in the exposure assessment. Similarly, if exposures are calculated for a given calendar year (e.g., 1975), individuals of different ages will be included in the assessment. As another example, if exposures are calculated for individuals 10 years prior to diagnosis/recruitment, individuals of different ages and individuals diagnosed in different calendar years will be included in the assessment. Therefore, these three temporal orientations can result in unique temporally specific exposures that reveal different aspects of underlying associations between disease risk and the timing of exposure over the life-course.

As an example of how alternative temporal orientations can influence the apparent timing of exposure, individual lifetime exposure to arsenic is displayed for three participants using years prior to diagnosis/interview and participant's age as the temporal orientation (Figure 5.4). Even using only these three participants, this figure indicates that the pattern of exposure among participants changes based on the temporal orientation. For example, Participants 2 and 3 both experience exposure > 50 µg/day during their first six years prior to diagnosis/interview; however, they do not experience these exposures at similar ages in their life. Therefore, when one uses age as the temporal orientation, their high exposures do not overlap. As another example, Participants 1 and 2 both experience elevated exposures when they are in their mid-forties, yet their exposure trajectories do not overlap when one uses years prior to diagnosis/interview as the temporal orientation. Among an entire study sample population, the trends of exposure between cases (diseased individuals) and controls (healthy individuals) will clearly shift when using different temporal orientations. Redefining exposure using alternative temporal orientations is most valuable in situations where a priori hypotheses concerning the timing of the exposure-disease relationship do not exist, as often is the case. This functionality allows health researchers to investigate whether exposure at any point in time, using multiple temporal orientations, is associated with development of disease. Factoring in human mobility and multiple temporal orientations may, therefore, be valuable for investigating the exposure-disease relationship in chronic diseases with long induction periods. Converting datasets between different temporal orientations is possible using space-time data structures in Time-GIS.

VALIDATE LIFETIME EXPOSURE AND QUANTIFY UNCERTAINTY IN THE EXPOSURE ESTIMATE

As the arsenic example illustrates, the ease with which space-time variability can be incorporated into detailed exposure assessments using Time-GIS has the potential to further the field of exposure science. One concern about Time-GIS-based exposure assessments, however, is the inability to completely validate past exposures. Not unlike other retrospective exposure assessment techniques, Time-GIS-based exposure estimates cannot be thoroughly validated, because gold standards do not exist for past exposures. Aspects of the exposure, however, could be validated using Time-GIS. For example, when assessing proximity to farmland, a proximity metric can first be calculated based on the intersection of a space-time raster dataset of land

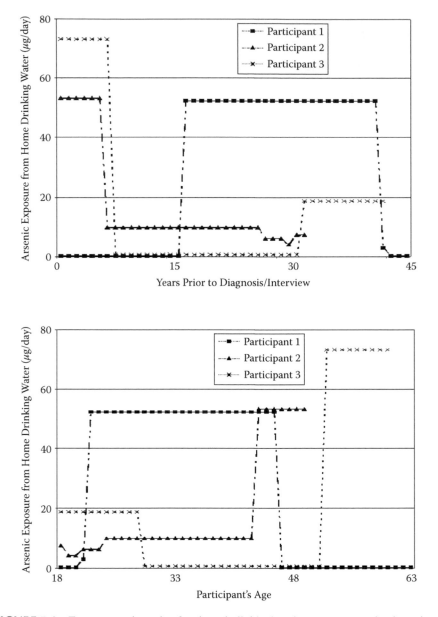

FIGURE 5.4 Exposure trajectories for three individuals using two temporal orientations: years prior to diagnosis/interview (top) and participant's age (bottom). For these three participants, their exposures appear to change relative to time and relative to each other when using different temporal orientations. Each temporal orientation can be used to reveal distinct insights into the temporal dynamics of the exposure-disease relationship.

use with a space-time vector dataset of residential history. This proximity metric could then be compared with a self-reported estimate of proximity to farmland at current and past residences. This comparison should shed light on how well people recall their proximity to farmland and how the accuracy of recall changes over time. In another example, space-time estimates of recent exposure to air pollutants could be compared with (1) spatial-only estimates of air pollution and (2) personal air filters from participant's breathing zones. This comparison would allow exposure scientists to quantify the added benefit from incorporating space-time variability in the exposure assessment. In order to conduct these types of comparisons, algorithms for calculating temporally specific measures of sensitivity, specificity, positive and negative predictive value, kappa statistics, and correlation need to be available in Time-GIS. These are familiar measures for validating models of continuous and categorical datasets in epidemiologic studies. Similar to results from the space-time join method or the exposure calculator, the results from these statistical measures will change with time in a temporally continuous fashion.

Along with tools for validating elements of the exposure assessment, algorithms for quantifying uncertainty may also prove valuable. Uncertainty may arise at several levels, including space-time estimates of environmental contaminants, recall and geocoding of mobility histories, and recall of temporally varying behaviors such as water consumption or showering frequency, among others. Broadly speaking, uncertainty in space-time exposure assessment is attributable to two sources: locational uncertainty and attribute uncertainty. Techniques for modeling both locational uncertainty and attribute uncertainty are well known and locational and attribute error models as described by Jacquez (1999) should be featured in Time-GIS. In addition, techniques for propagating the uncertainty in the parameter estimates up through the exposure metric need to be developed, perhaps using Latin-Hypercube sampling of probability distributions and Monte-Carlo simulation (Cox 1996; Viscarra Rossel et al. 2001). By using these approaches, the output of the exposure function will no longer be a single exposure value, but a set of possible values with a corresponding probability of occurrence (Goovaerts 2001). These techniques are necessary to ensure wide adoption of Time-GIS by the exposure science community.

CONCLUSIONS

Individuals and environmental contaminants change locations in time, and therefore tools that integrate space-time maps are necessary to more reliably calculate individual exposure. Time-GIS technology has recently been developed (AvRuskin et al. 2004; Greiling et al. 2005; Jacquez et al. 2005; Meliker et al. 2005; Weaver et al., in press) and has the potential to advance the assessment of individual-level exposure to environmental contaminants. The first application of Time-GIS for exposure assessment was recently published in a bladder cancer case-control study of lifetime exposure to arsenic in drinking water (Meliker et al. 2007). In that example, spatially, temporally, and spatiotemporally varying databases were seamlessly integrated, including residential mobility, occupational mobility, changes in locations of public and private water supplies, changes in arsenic concentration in public water supplies, changes in water consumption, and changes in food consumption. While that example clearly highlighted the potential of Time-GIS for exposure reconstruction, in this chapter I have described and illustrated several additional functionalities. By designing Time-GIS technology to easily integrate data across space-time vector and space-time raster maps, import space-time datasets from models of environmental contaminants, calculate estimates of lifetime exposure, incorporate different temporal orientations for calculating exposure, and quantify uncertainty in the lifetime exposure assessment, exposure scientists will no longer be forced to simplify the space-time variability inherent in their datasets.

Time-enabled geographic approaches explicitly characterize the timing and ordering of exposure, accounting for temporal changes in the magnitude and locations of exposure. By efficiently and easily incorporating space-time variability, Time-GIS enables flexibility in the assessment of temporally detailed individual-level exposure estimates. This flexibility in integrating information across space-time datasets allows the user to calculate a series of alternative exposure metrics and to refine the exposure estimates as the underlying models of environmental contamination are improved. This type of iterative process is impractical with traditional GIS because of the time and labor involved in generating an exposure estimate. These improved estimates should be valuable for exposure assessment in environmental epidemiological studies and for risk assessment, surpassing traditional GIS-based efforts that are less reliable and less accurate because they ignore the spatiotemporal variability inherent in many datasets. A clear opportunity exists for integrating space-time data but, until recently, it has not been clear how this might be accomplished. The Time-GIS approach constitutes a valuable tool for advancing traditional means of exposure reconstruction.

REFERENCES

Aschengrau, A., D. P. Ozonoff, R. Coogan, T. Vezina, T. Heeren, and Y. Zhang. 1996. Cancer risk and residential proximity to cranberry cultivation in Massachusetts. *Am J Public Health* 86:1289–96.

AvRuskin, G. A., G. M. Jacquez, J. R. Meliker, M. J. Slotnick, A. M. Kaufmann, and J. O. Nriagu. 2004. Visualization and exploratory analysis of epidemiologic data using a novel space time information system. *Int J Health Geogr* 3:26.

Baker, M. N. and M. Taras. 1981. The Quest for Pure Water: The History of Water Purification from the Earliest Records to the Twentieth Century. New York: American Water Works Association.

Barnes, S. and A. Peck. 1994. Mapping the future of health care: GIS applications in health care analysis. *Geographic Information Systems* 4:31–33.

Beaubroef, T. and J. Breckenridge. 2000. Real-world issues and applications for real-time geographic information systems (RT-GIS). *J Navigation* 53:124–31.

Bellander, T., N. Berglin, P. Gustavsson, T. Jonson, F. Nyberg, G. Pershagen, and L. Jarup. 2001. Using geographic information systems to assess individual historical exposure to air pollution from traffic and house heating in Stockholm. *Environ Health Perspect* 109:633–39.

Beyea, J. and M. Hatch. 1999. Geographic exposure modeling: A valuable extension of geographic information systems for use in environmental epidemiology. *Environ Health Perspect* 107 (suppl 1): 181–90.

Bonner, M. R., D. Han, J. Nie, P. Rogerson, J. E. Vena, P. Muti, M. Trevisan, S. B. Edge, and J. L. Freudenheim. 2005. Breast cancer risk and exposure in early life to polycyclic aromatic hydrocarbons using total suspended particulates as a proxy measure. *Cancer Epidem Biomar* 14:53–60.

Brody, J. G., D. J. Vorhees, S .J. Melly, S. R. Swedis, P. J. Drivas, and R. A. Rudel. 2002. Using GIS and historical records to reconstruct residential exposure to large-scale pesticide application. *J Expo Anal Env Epid* 12:64–80.

Brownstein, J. S., T. R. Holford, and D. Fish. 2003. A Climate-Based Model Predicts the Spatial Distribution of the Lyme Disease Vector Ixodes scapularis in the United States. *Environ Health Perspect* 111:1152–57.

Clarke, K. C., S. L. McLafferty, and B. J. Tempalski. 1996. On epidemiology and geographic information systems: A review and discussion of future directions. *Emerg Infect Dis* 2:85–92.

Cox, L. A. 1996. Reassessing benzene risks using internal doses and Monte-Carlo uncertainty analysis. *Environ Health Perspect* 104 (Suppl 6):1413–29.

Croner, C. M., J. Sperling, and F. R. Broome. 1996. Geographic Information Systems (GIS): New perspectives in understanding human health and environmental relationships. *Stat Med* 15:1961–1977.

Dragicevic, S. and D. J. Marceau 2000. A fuzzy set approach for modeling time in GIS. *Int J Geog Inf Sci* 14:225–45.

Elgethun, K., R. A. Fenske, M. G. Yost, and G. J. Oalcisko. 2003. Time-location analysis for exposure assessment studies of children using a novel global positioning system instrument. *Environ Health Perspect* 111:115–22.

Elliott, P., D. Briggs, S. Morris, C. de Hoogh, C. Hurt, T. K. Jensen, I. Maitland, S. Richardson, J. Wakefield, and L. Jarup. 2001. Risk of adverse birth outcomes in populations living near landfill sites. *BMJ* 323:363–68.

Galton, A. 2003. Desiderata for a spatio-temporal geo-ontology. In COSIT 2003, ed. W. Kuhn, M. F. Worboys, S. Timpf, 1–12. *LNCS* 2825.

Goodchild, M. 2000. GIS and Transportation: Status and Challenges. *GeoInformatica* 4:127–39.

Goovaerts, P. 2001. Geostatistical modelling of uncertainty in soil science. *Geoderma* 103:3–26.

Goovaerts, P., G. AvRuskin, J. Meliker, M. Slotnick, G. M. Jacquez, and J. Nriagu. 2005. Geostatistical modeling of the spatial variability of arsenic in groundwater of Southeast Michigan. *Water Resour Res* 41(7):W07013 10.1029.

Greiling, D.A., G. M. Jacquez, A. M. Kaufmann, and R. G. Rommel. 2005. Space time visualization and analysis in the Cancer Atlas Viewer. *Journal of Geographical Systems* 7:67–84.

Hagerstrand, T. 1970. What about people in regional science? Papers of the Regional Science Association 24:7–21.

Hornsby, K. and M. Egenhofer. 2000. Identity-based change: a foundation for spatio-temporal knowledge representation. *Int J Geog Inf Sci* 14:207–24.

Hornsby, K. and M. Egenhofe. 2002. Modeling moving objects over multiple granularities. *Ann Math Artif Intel* 36:177–94.

Jacquez, G. M. 1999. Spatial statistics when locations are uncertain. *Geographic Information Sciences* 5:77–87.

Jacquez, G. M. 2004. Current practices in the spatial analysis of cancer: flies in the ointment. *Int J Health Geogr* 3:22.

Jacquez, G. M., D. Greiling, and A. Kaufmann. 2005. Design and Implementation of Space Time Information Systems. *Journal of Geographical Systems* 7:7–24.

Mark, D. M. and M. J. Egenhofer. 1998. Geospatial Lifelines. In *Integrating Spatial and Temporal Database,* ed. O. Guenther, T. Sellis, B. Theodoulidis, 14. Dagstuhl Seminar Report no. 228.

Mark, D., M. J. Egenhofer, L. Bian, K. E. Hornsby, P. A. Rogerson, and J. Vena. 2000. Spatiotemporal GIS analysis for environmental health: Solutions using geospatial lifelines. In *Geography and medicine,* ed. A. Flahaut, L. Toubiana, A. J. Valleron, 65–78. GEOMED '99. Paris: Elsevier.

Mather, F. J., L. C. Whited, E. C. Langlois, C. F. Shorter, C. M. Swalm, J. G. Shaffer, and W. R. Harley. 2004. Statistical methods for linking health, exposure and hazards. *Environ Health Perspect* 112:1440–45.

McNamee, R. and H. Dolk. 2001. Does exposure to landfill waste harm the fetus? *BMJ* 323:351–52.

Meliker, J. R., M. J. Slotnick, G. A. AvRuskin, A. Kaufmann, G. M. Jacquez, and J. O. Nriagu. 2005. Improving exposure assessment in environmental epidemiology: Application of spatio-temporal visualization tools. *Journal of Geographical Systems* 7:49–66.

Meliker, J. R., M. J. Slotnick, G. A. AvRuskin, A. Kaufmann, S. A. Fedewa, P. Goovaerts, G. M. Jacquez, and J. O. Nriagu 2007. Individual lifetime exposure to inorganic arsenic using a Space-Time Information System. *Int Arch Occ Env Health* 80:184–97.

Nuckols, J. R., M. H. Ward, and L. Jarup. 2004. Using Geographic Information Systems for Exposure Assessment in Environmental Epidemiology Studies. *Environ Health Perspect* 112:1007–15.

Nyberg, F., P. Gustavsson, L. Jarup, T. Bellander, N. Berglind, R. Jakobsson, and G. Pershagen. 2000. Urban air pollution and lung cancer in Stockholm. *Epidemiology* 11:487–95.

O'Leary, E. S., J. E. Vena, J. L. Freudenheim, and J. Brasure. 2004. Pesticide Exposure and Risk of Breast Cancer: A Nested Case-Control Study of Residentially Stable Women Living on Long Island. *Environ Res* 94:134–44.

Peuquet, D. J. 2001. Making space for time: Issues in space-time data representation. *Geoinformatica* 5:11–32.

Phillips, M. L., T. A. Hall, N. A. Esmen, R. Lynch, and D. L. Johnson. 2001. Use of global positioning system technology to track subject's location during environmental exposure sampling. *J Expo Anal Env Epid* 11:207–15.

Pickle, L., L. Waller, and A. B. Lawson. 2005. Current Practices in cancer spatial data analysis: a call for guidance. *Int J Health Geogr* 4:3.

Pickle, L. W., M. Szczur, D. R. Lewis, and D. G. Stinchcomb. 2006. The crossroads of GIS and health information: A workshop on developing a research agenda to improve cancer control. *Int J Health Geog* 5:51.

Reif, J. S., J. B. Burch, J. R. Nuckols, L. Metzger, D. Ellington, and W. K. Anger. 2003. Neurobehavioral Effects of Exposure to Trichloroethylene Through a Municipal Water Supply. *Environ Res* 93:248–58.

Reynolds, P., S. E. Hurley, R. B. Gunier, S. Yerabati, T. Quach, and A. Hertz. 2005. Residential proximity to agricultural pesticide use and incidence of breast cancer in California, 1988–1997. *Environ Health Perspect* 113:993–1000.

Rodriguez, A., N. van de Weghe, and P. De Maeyer. 2004. Simplifying sets of events by selecting temporal relations. In GIScience 2004, ed. M. J. Egenhofer, C. Freksa, H. J. Miller. *LNCS* 3234:269–84.

Ross, Z., P. B. English, R. Scalf, R. Gunier, S. Smorodinsky, S. Wall, and M. Jerrett. 2006. Nitrogen dioxide prediction in Southern California using land use regression modeling: Potential for environmental health analyses. *J Expo Sci Env Epid* 16:106–14.

Rull, R. P., B. Ritz, and G. M. Shaw. 2006. Validation of self-reported proximity to agricultural crops in a case–control study of neural tube defects. *J Expo Sci Env Epid* 16:147–55.

Sinha, G. and D. Mark. 2005. Measuring similarity between geospatial lifelines in studies of environmental health. *Journal of Geographical Systems* 7:115–36.

Stellman, J. M., S. D. Stellman, T. Weber, C. Tomasallo, A. B. Stellman, and R. Christian. 2003. A Geographic Information System for Characterizing Exposure to Agent Orange and Other Herbicides in Vietnam. *Environ Health Perspect* 111:321–28.

Storm, H. H., E. V. Michelsen, I. H. Clemmensen, and J. Pihl. 1997. The Danish Cancer Registry — history, content, quality, and use. *Dan Med Bull* 44:549–53.

Swartz, C. H., R. A. Rudel, J. R. Kachajian, and J. G. Brody. 2003. Historical Reconstruction of Wastewater and Land Use Impacts to Groundwater Used for Public Drinking Water: Exposure Assessment Using Chemical Data and GIS. *J Expo Anal Env Epid* 13:403–16.

Viscarra Rossel, R. A., P. Goovaerts, and A. B. McBratney. 2001. Assessment of the production and economic risks of site-specific liming using geostatistical uncertainty modeling. *Environmetrics* 12:699–711.

Ward, M. H., J. R. Nuckols, S. J. Weigel, S. K. Maxwell, K. P. Cantor, and R. S. Miller. 2000. Identifying populations potentially exposed to agricultural pesticides using remote sensing and a geographic information system. *Environ Health Perspect* 108:5–12.

Weaver, C., D. Fyfe, A. Robinson, D. Holdsworth, D. Peuquet, and A. M. MacEachren. In Press. Visual Analysis of Historic Hotel Visits. *Information Visualization*.

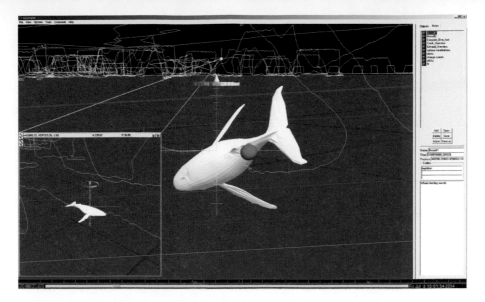

FIGURE 1.4 The GeoZui4D interface is shown being used to analyze whale behavior. The buttons on the right-hand side are space-time notes. They capture a view, a time, and an annotation.

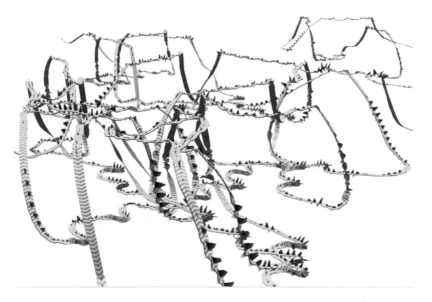

FIGURE 1.5 This track of a humpback whale shows the attitude of the whale over a period of a few hours. The sawtooths indicate where fluking occurred. The yellow portions of the track show where the roll angle exceeded 40 degrees from horizontal. It is immediately evident that a highly stereotyped behavior is occurring at the bottom of the dives where the animal repeatedly rolls on its side for a fixed duration before righting itself.

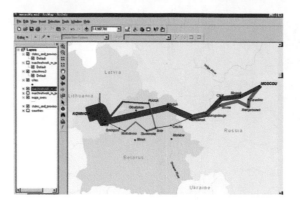

FIGURE 2.3   A re-rendering of the Minard map of Napoleon's 1812 Moscow campaign.

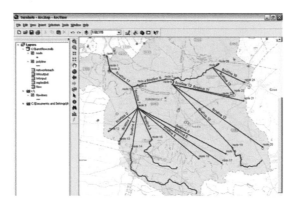

FIGURE 2.4   The hydrology of a part of the Central Kentucky Karst. Red lines indicate inferred flows.

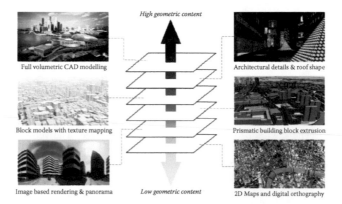

FIGURE 8.1   A typology of 3D urban models using different modeling methods (reproduced from Shiode, 2001).

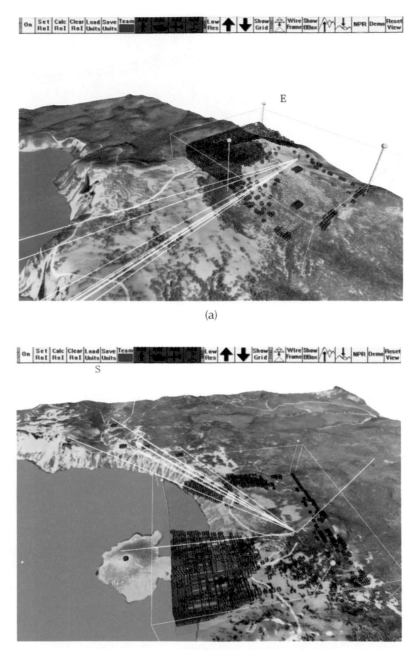

FIGURE 9.4 Example results of both point-to-point and point-to-volume visibility calculations in our LoS application. Volumetric results are depicted as black boxes in a yellow region-of-interest. (Note that the voxel sizes here have been enlarged for illustrative purposes.) Point-to-point visibility from the red objects are shown here as connecting lines-of-sight, but can also be represented as icons above visible or invisible units.

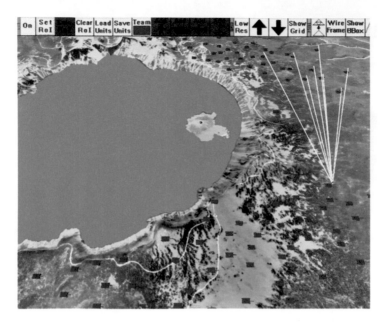

FIGURE 9.5 Line-of-sight scenario consisting of two teams of 53 units each across a 20 km × 14 k terrain. Statistics for this scenario at different levels of accuracy are given in Table 1.

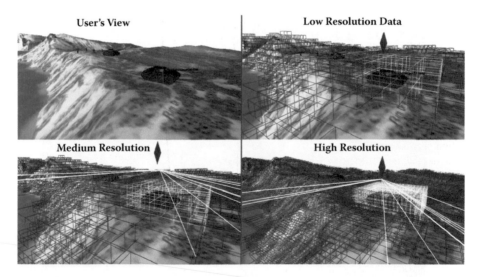

FIGURE 9.6 The user's view presents a simplified terrain mesh, while underneath the calculations are performed on the volumetric models. The application can adaptively switch between multiple resolutions (each with their own confidence measures), maintaining the desired balance between computational speed and accuracy or confidence of the results.

FIGURE 9.7 Excerpt from testbed dataset with collection of building models, underlying DEM, and classified LIDAR point cloud.

FIGURE 9.9 Outline of an urban model after aggregation. The left image shows the original outline of the aggregated models; the right shows the simplified outline. The simplification process preserves the legibility elements in the city model, thereby creating a simplified model that remains legible and understandable.

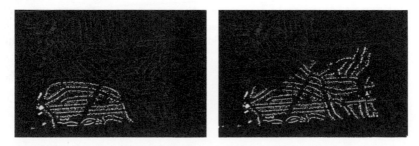

FIGURE 9.11 Clustering buildings in a city. The left image shows clustering results that follow the urban legibility element Paths. The right image shows the result of a more traditional distance based clustering.

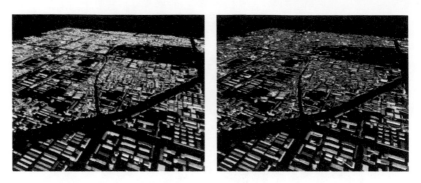

FIGURE 9.12 Original textured 3D model (left) of Xinxiang, China; simplified model (right) with only 18 percent of the original number of polygons and aggregated textures. View-dependent rendering is applied to the hierarchical multiresolution structure on the right.

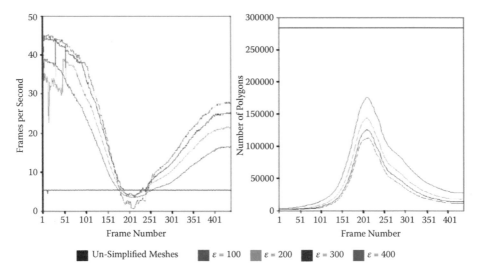

FIGURE 9.13 Frame rate and polygon counts. Using simplified models with different levels of detail in a flythrough scene (a high ε value denotes a large amount of simplification), we see that our simplification method can provide drastic speedup by decreasing the number of polygons rendered to the screen.

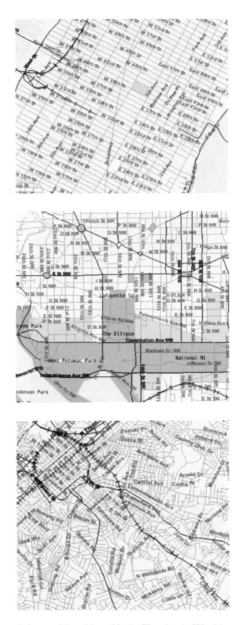

FIGURE 9.15 Views of three cities: New York City (top), Washington, D.C. (middle), and Charlotte (bottom). These three cities have distinctively different layouts; New York City resembles a gridlike structure; Washington, D.C., is radial with roads emanating from the Congress and the White House; and Charlotte is mostly unstructured with strong "sprawling" feeling.

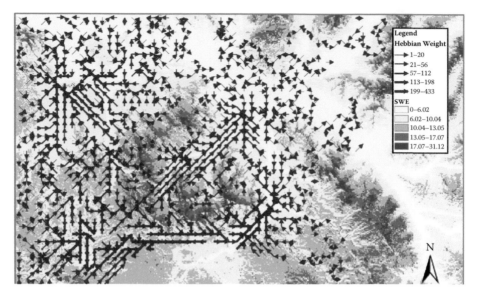

FIGURE 10.7   Migration routes produced by Hebbian learning.

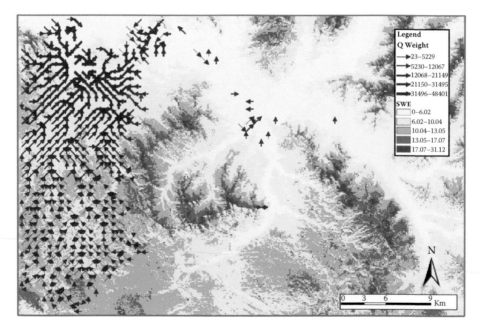

FIGURE 10.8   Migration routes produced by Q learning.

6 Analysis of Human Space-Time Behavior: Geovisualization and Geocomputational Approaches

Mei-Po Kwan and Fang Ren

CONTENTS

THE ANALYSIS OF HUMAN SPACE-TIME BEHAVIOR

The study of human activities and movements in space and time in the urban context has long been an important theme in social science research. It covers a wide range of topics, such as migration, residential mobility, shopping, travel, and commuting behavior. A major difficulty in the analysis of human space-time behavior (HSTB) is that individual movement in space-time is a complex trajectory with many interacting dimensions. These include the location, timing, duration, sequencing, and type of activities and trips. This characteristic of HSTB has made the simultaneous analysis of its many dimensions difficult. Two different approaches were adopted to resolve this problem in past research. On one hand, some studies focused on a few component dimensions of HSTB at a time (e.g., Golob and McNally 1997; Goulias 1999). On the other hand, some studies treat HSTB as a multidimensional whole and use multivariate methods to derive generalized behavioral patterns from a large number of variables (e.g., Bhat and Singh 2000; Golob 1985; Ma and Goulias 1997a, b; Recker et al. 1987; Shoval and Isaacson 2007).

The development and application of these quantitative methods have enhanced our understanding of HSTB in significant ways. For instance, through the use of multivariate group identification methods, such as clustering or pattern recognition algorithms, complex patterns in the original data set can be represented by some general characteristics and organized into a relatively small number of homogenous classes. Further, once patterns in HSTB are represented in terms of a limited number of categories, they can be related to a large number of attributes of the individuals or households that generate them and used as a response variable in models of HSTB. While these quantitative or statistical methods are useful for modeling purposes and for discovering the complex interrelations among variables, they also have their limitations.

First, since many statistical methods used in past studies (e.g., log-linear models) are designed to deal with categorical data, organizing the original data in terms of discrete units of space and time has been a necessary step in most analyses of HSTB in the past. Discretization of temporal variables, such as the start time or duration of activities, involves dividing the relevant span of time into several units and assigning each activity or trip into the appropriate class (e.g., dividing a day into 8 or 12 temporal divisions into which activities or trips are grouped). Discretization of spatial variables, such as distance from home, involves dividing the relevant distance range into several "rings." Since both the spatial and temporal dimensions are continuous, results of any analysis that are based on these discretized variables may be affected by the particular schema of spatial and temporal divisions used. The problem may be serious when dealing with the interaction between spatial and temporal variables, since two discretized variables are involved.

Second, few of these methods were designed to handle real geographical locations of HSTB in the context of a study area. Often, the spatial dimension is represented by some measures derived from real geographical locations (e.g., distance or direction from a reference point such as home or workplace of an individual). Further, locational information of activities or trips was often aggregated with respect to a zonal division of the study area (e.g., traffic analysis zones or census tracts). Using such zone-based data, measurement of location and/or distance involves using zone centroids where information about specific activity locations in geographic space and their spatial relations with other urban opportunities is lost (Kwan and Hong 1998). Third, as detailed data about HSTB have become more readily available in recent years, effective methods for exploring these data are also urgently needed (McCormack 1999). Without them, the researcher may need to model HSTB without a preliminary understanding of the behavioral characteristics or uniqueness of the individuals in the sample at hand. This can be costly in later stages of a study if the model's specifications fail to take into account the behavioral anomalies involved.

Conceptual frameworks and analytical methods that would help the researcher to effectively overcome these three difficulties in the analysis of HSTB are needed. In this chapter we present the time-geographic perspective as a framework that provides a sound conceptual basis for a range of analytical methods that overcome some of these difficulties. We then discuss some of the GIS-based time-geographic methods that were developed in recent years. These include three-dimensional (3D) geovisualization and geocomputational methods. Usefulness of these methods is illustrated through examples drawn from recent studies by us and other researchers. We also attempt to show that GIS provides an effective environment for implementing time-geographic constructs and for the future development of operational methods in time-geographic research. We suggest that GIS-based time-geographic methods are effective means for the study of human activities and movements in space and time in the urban context.

TIME GEOGRAPHY

As discussed in the last section, an effective conceptual framework is needed for mitigating the difficulties in the analysis of HSTB. First, we need a conceptual framework for developing methods that avoids the discretization of the spatial and temporal variables. Second, the framework should be attentive to and could provide a basis for taking real geographical locations of HSTB into account. Third, these methods should facilitate exploratory spatial data analysis (ESDA), which helps to make modeling efforts more focused and fruitful in later stages of a study. We propose that time geography offers such a conceptual framework for developing methods for the analysis of HSTB.

Time geography was developed by a group of Swedish geographers at Lund University in the 1950s to '60s — including Torsten Hägerstrand, Tommy Carlstein, Bo Lenntorp, and Don Parkes (Kwan 2004). Important constructs in time geography, such as stations, projects, space-time paths, and prism constraints are well articulated in Carlstein et al. (1978), Hägerstrand (1970), Parkes and Thrift (1975), and Thrift (1977). Lenntorp (1976) provided the first in-depth analytical treatment and operationalization of time-geographic constructs.

Time geography conceives an individual's activities and travel in a twenty-four-hour day as a continuous temporal sequence in geographical space. The trajectory that traces this activity sequence is referred to as a space-time path, while the graphical representation of the three-dimensional space in which this path unfolds is referred to as the space-time aquarium. The number and location of everyday activities that can be performed by a person are limited by the amount of time available and the space-time constraints associated with various obligatory activities (e.g., work) and joint activities with others. These constraints largely arise from the spatial or temporal rigidity associated with certain types of activities people undertake in their daily lives. These activities are called fixed activities (for example, working or visiting a doctor) because it is difficult to change the place or time to perform them, and as a result they also tend to restrict a person's freedom to undertake other spatially and temporally flexible activities. The space-time constraints imposed by fixed activities are collectively referred to as fixity constraint. This time-geographic conception is valuable for understanding HSTB because it integrates the temporal and spatial dimensions of human-activity patterns into a single analytical framework. Although time, in addition to space, is a significant element in structuring individual activity patterns, past approaches mainly focus on either their spatial or temporal dimension. The significance of the interaction between the spatial and temporal dimensions in structuring individual daily space-time trajectories is often ignored.

Time geography not only highlights the importance of space for understanding the geographies of everyday life. It also allows the researcher to examine the complex interaction between space and time and their joint effect on the structure of HSTB in particular places (Cullen et al. 1972). It can be applied in a wide range of fields and research areas. Since the early 1990s, the perspective has been particularly useful for understanding women's activity-travel behavior, because it helps to identify the restrictive effect of space-time constraints on their activity choice, job location, travel, and occupational and employment status (Kwan 1999a, b, 2000a; Laws 1997; Tivers 1985). Time geography has also been used as a framework for the study of migration and mobility behavior (Odland 1998), exposure to health risk, and the everyday life of the elderly, children, and homeless people (e.g., Mårtensson 1977; Rollinson 1998). Many transportation geographers and researchers have also found the time-geographic perspective useful for modeling human activity-travel behavior (e.g., Kim and Kwan 2003; Miller and Wu 2000; Yu and Shaw 2004).

Despite the usefulness of time geography in many areas of social science research, very few studies have actually implemented its constructs as analytical methods up to the mid-1990s — with the notable exception of Bo Lenntorp's Program Evaluating the Set of Alternative Sample Path (PESASP) simulation model. The limited development of time-geographic methods was largely due to the lack of detailed geographical and individual-level data as well as analytical tools that can realistically represent the complexities of an urban environment (e.g., the transportation network and spatial distribution of urban opportunities). For example, a study that dealt with 286 urban opportunities spent about three months in the manual construction of a digital street network that only had 939 nodes and 2,395 arcs in the late 1970s. So the time needed for constructing the geographic data for time-geographic studies had been considerable. Another difficulty is that the algorithms used to implement time-geographic methods have been computationally intensive.

However, with increasing availability of digital geographic databases of urban areas and georeferenced individual-level data, as well as improvement in the representational and geocomputational capabilities of Geographical Information Systems (GIS), it is now more feasible to operationalize and implement time-geographic constructs (e.g., Frihida et al. 2004; Hendricks et al. 2003). Further, the use of GIS also allows the incorporation of large amounts of geographic data that are essential for any meaningful analysis of HSTB. Because of these changes, time-geographic methods are undergoing a new phase of development as several recent studies indicate (Kwan 1998, 2000b; Miller 1991, Ohmori et al. 1999; Takeda 1998; Weber and Kwan 2002). Although the primary focus of these studies is on individual accessibility, time geography can be fruitfully applied in many areas of social science research.

GIS-BASED 3D GEOVISUALIZATION METHODS

Two areas in which time-geographic methods have been used in the analysis of HSTB in recent years are GIS-based three-dimensional (3D) geovisualization and geocomputational methods. Geovisualization is the use of concrete visual representations and human visual abilities to make spatial contexts and problems visible. Through involving the geographical dimension in the visualization process, it greatly facilitates the identification and interpretation of spatial patterns and relationships in complex data in the geographical context of a particular study area.

The use of GIS-based 3D geovisualization in time-geographic research is a rather recent phenomenon. In most early studies, 2D maps and graphical methods were used to portray HSTB (e.g., Chapin 1974; Tivers 1985). Individual daily space-time paths were represented as lines connecting various destinations. Using such kinds of 2D graphical methods, information about the timing, duration, and sequence of activities and trips was lost. Recent years, however, have seen noticeable change. As more georeferenced activity-travel diary data become available, and as more GIS software has incorporated 3D capabilities, GIS-based 3D geovisualization has become a more feasible approach for time-geographic research.

For instance, Forer (1998) implemented space-time paths and prism on a 3D raster data structure for visualizaton and computational purposes. Their method is useful for aggregating individuals with similar socioeconomic characteristics and for identifying behavioral patterns. Kwan (2000b) and Kwan and Lee (2004) implemented 3D geovisualization of space-time paths and aquariums using vector GIS methods and activity-travel diary data. These studies indicate that GIS-based geovisualization can be a fruitful method for time-geographic research. Further, implementing 3D visualization of HSTB can be an important first step in the development of GIS-based geocomputational procedures that are applicable in many areas of social-science research. The following subsections provide an overview of some of these methods from our recent research.

A note on the quality of the figures is warranted before presenting the geovisualization results in the following subsections. Since the 3D patterns involved are highly complex and the figures are non-scalable raster images produced from screen captures, there are many limitations on producing clear illustrations using 2D graphics. The reader may find it hard to follow the discussion simply by looking at these figures, since the text is based on observations enabled by the computer-aided interactive 3D geovisualization environment (which is not available to the reader). The difficulty the reader may have in "seeing" these 2D images clearly is the inevitable outcome of the need to present results of the color display of complex 3D patterns in the form of 2D graphics. The visual quality of the 2D figures in the paper, therefore, should not undermine the argument that interactive 3D geovisualization is hepful for understanding HSTB. Further, as these figures cannot convey the same amount and quality of information enabled only by interactive 3D geovisualizations, it is best to treat them as illustrations of what one might see when performing the interactive 3D geovisualizations. Their purpose is to give the reader a feel for what the interactive geovisualization does. To fully appreciate the value of the methods discussed in the paper, one needs to go through the real computer-aided interactive geovisualization sessions instead of looking at their black-and-white 2D representations.

SPACE-TIME PATHS AND THE SPACE-TIME AQUARIUM

The earliest 3D method for the visualization of individual space-time paths is the space-time aquarium conceived by Hägerstrand (1970). In a schematic representation of the aquarium, the vertical axis is the time of day, and the boundary of the horizontal plane represents the spatial scope of the study area. Individual space-time paths are portrayed as trajectories in this 3D aquarium. Although the schematic representation of the space-time aquarium was developed long ago, it has never been implemented using real activity-travel diary data. The main difficulties include the need to convert the activity data into "3Dable" formats that can be used by existing visualization software, and the lack of comprehensive geographic data for representing complex geographic objects of the urban environment.

Kwan (1999a) was the first study that implemented the space-time aquarium and space-time paths in a 3D GIS environment using individual-level activity-travel diary data. Based on a subsample of seventy-two European Americans from the data set she collected in Columbus, Ohio, the study examines the effect of gender on the space-time patterns of out-of-home nonemployment activities. Visualization of the space-time paths of three groups of research participants reveals their distinctive activity patterns, and this insight guided the structural equation modeling in the later phase of the study. The results of the study show that the structure of one's daily activity patterns and daytime fixity constraint depends more on one's gender than on some conventional variables of household responsibilities such as the presence or number of children in the household.

Given the limited 3D visualization capabilities of GIS when the study was conducted, the geovisualization performed in Kwan (1999a) did not use a large geographic database of the study area (but it did incorporate a transportation network). The first study that incorporated a comprehensive geographic database for the 3D geovisualization of space-time paths was that of Kwan (2000b). This study used the Portland activity-travel diary data set mentioned earlier and a comprehensive geographic database of the Portland, Oregon, metropolitan region. The GIS database provides comprehensive data on many aspects of the urban environment and transportation system of the study area. It has data for about 400,000 land parcels in the study area. The digital street network used, with 130,141 arcs and 104,048 nodes, covers the four counties of the study area (i.e., Clark, Clackamas, Multnomah, and Washington).

These contextual data allow the activity-travel data to be related to the geographical environment of the region during visualization. To implement 3D geovisualization of the space-time aquarium, four contextual geographic data layers are first converted from 2D map layers to 3D format and added to a 3D scene. These include the metropolitan boundary, freeways, major arterials, and rivers. For better close-up visualization and for improving the realism of the scene, outlines of commercial and industrial parcels in the study area are converted to 3D polygons and vertically extruded in the scene. Finally, the 3D space-time paths of the African- and

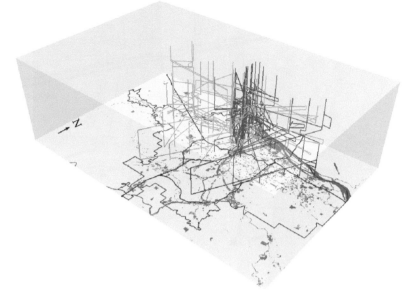

FIGURE 6.1 Space-time aquarium showing the space-time paths of African- and Asian-Americans in the sample. (From Kwan, M.-P. and J. Lee, Geovisualization of human activity patterns using 3D GIS: A time-geographic approach. In Spatially Integrated Social Science, ed. M. F. Goodchild and D. G. Janelle, 48–66. New York: Oxford University Press, 2004. Use by permission of Oxford University Press, Inc.)

Asian-Americans in the sample are generated and added to the 3D scene. These procedures finally created the scene shown in Figure 6.1.

The overall pattern of the space-time paths for these two groups shown in Figure 6.1 indicates heavy concentration of daytime activities in and around downtown Portland. Using the interactive visualization capabilities of the 3D GIS, it was observed that many individuals of these two ethnic groups work in downtown Portland and undertake a considerable amount of their nonemployment activities in areas within and east of the area. Space-time paths for individuals who undertook several nonemployment activities in a sequence within a single day tend to be more fragmented than those who have long work hours during the day. Further, ethnic differences in the spatial distribution of workplace are observed using the interactive capabilities provided by the geovisualization environment. The space-time paths of Asian-Americans are more spatially scattered throughout the area than those of the African-Americans, whose work and nonemployment activities are largely concentrated in the east side of the metropolitan region.

A close-up view from the west of the 3D scene is given in Figure 6.2, which shows some of the details of downtown Portland in areas within and around the "loop" and along the Willamette River. Portions of some space-time paths can also be seen in the figure as well. With the 3D parcels and other contextual layers in view, the figure gives the researcher a strong sense about the geographical context through a virtual realitylike view of the downtown area.

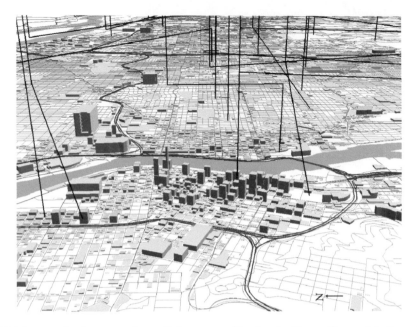

FIGURE 6.2   A close-up view of downtown Portland. (From Kwan, M.-P. and J. Lee, Geovisualization of human activity patterns using 3D GIS: A time-geographic approach. In Spatially Integrated Social Science, ed. M. F. Goodchild and D. G. Janelle, 48–66. New York: Oxford University Press, 2004. Use by permission of Oxford University Press, Inc.)

Space-Time Paths Based on GPS Data

Although the 3D space-time paths shown in Figures 6.1 and 6.2 are helpful for understanding the activity patterns of different population subgroups, these paths are not entirely realistic since they only connect trip ends with straight lines and do not trace the travel routes of an individual. This limitation is due to the lack of route data in the Portland dataset. When georeferenced activity-travel data collected by GPS are available and used in the geovisualization environment, the researcher can examine the detailed characteristics of an individual's space-time behavior as actual-travel routes can be revealed by this kind of data (Kwan 2000c; Kwan and Lee 2004). Figure 6.4 illustrates this possibility using the GPS data collected in the Lexington Area Travel Data Collection Test conducted in 1997 (Battelle 1997). The original data set contains information of 216 licensed drivers (100 male, 116 female) from 100 households with an average age of 42.5. In total, data of 2,758 GPS-recorded trips and 794,861 data points of latitude-longitude pairs and time were collected for a six-day period for each survey participant.

To prepare for 3D geovisualization, three contextual geographic data layers of the Lexington metropolitan area are first converted from 2D map layers to 3D format and added to a 3D scene. These include the boundary of the Lexington metropolitan region, highways, and major arterials. As an illustration, the 3D space-time paths of the women without children under 16 years of age in the sample are generated

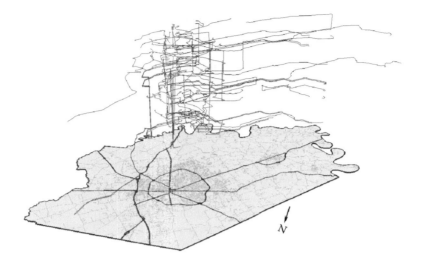

FIGURE 6.3 Space-time paths based on GPS data collected in Lexington, Kentucky. (From Kwan, M.-P. and J. Lee, Geovisualization of human activity patterns using 3D GIS: A time-geographic approach. In Spatially Integrated Social Science, ed. M. F. Goodchild and D. G. Janelle, 48–66. New York: Oxford University Press, 2004. Use by permission of Oxford University Press, Inc.)

and added to the 3D scene. These procedures finally created the scene shown in Figure 6.3. The overall pattern of the space-time paths for these women indicates that their trips were undertaken using largely highways and major arterials. There is some regularity as indicated by the daily repetition of trips in more or less the same time throughout the six-day survey period. This suggests that distinctive patterns of space-time behavior can be revealed by 3D geovisualization.

The computational intensity of processing and visualizing large space-time datasets causes difficulties in the analysis of these GPS data (Kwan 2001a; Lowe 2003). For instance, the original GPS data file for the 100 households contains 794,861 data points of latitude-longitude pairs and time (Kwan 2000c; Murakami and Wagner, 1999). It takes up about 230 megabytes of disk space in the format provided on the data CD. Manipulating files of this size can be taxing for the computer hardware normally available to social scientists. Although improvement in computing power in the near future will reduce this problem, much research is still needed to develop more efficient algorithms and data manipulation methods for handling large GPS datasets.

HUMAN EXTENSIBILITY AND CYBERSPATIAL ACTIVITIES IN SPACE-TIME

Three-dimensional geovisualization has also been applied to visualizing human activities in both the physical world and cyberspace based on the notion of human extensibility (Kwan 2000d). The concept of the individual as an extensible agent was first formulated by Janelle (1973), where extensibility represents the ability of a person to overcome the friction of distance through using space-adjusting technologies, such as transportation and communication. Human extensibility not only expands a person's scope of sensory access and knowledge acquisition, it also enables a person to engage in distantiated social actions whose effect may extend across disparate geographical regions or historical episodes (Kwan 2001b). To depict human extensibility that includes activities in both the physical world and cyberspace, Adams (1995) developed the extensibility diagram using the cartographic medium. The diagram, based on Hägerstrand's space-time aquarium, portrays a person's daily activities and interactions with others as multiple and branching space-time paths in three dimensions, where simultaneity and temporal disjuncture of different activities are revealed. This method can be used to represent a diverse range of human activities in both the physical and virtual worlds, including telephoning, driving, e-mailing, reading, remembering, meeting face-to-face, and television viewing.

Although the extensibility diagram is largely a cartographic device, most of its elements are amenable to GIS implementation. Kwan (2000d) developed a method for implementing the extensibility diagram using 3D GIS. The study used real data about a person's physical activities and cyberspatial activities (e.g., e-mail messages and Web-browsing sessions). The focus is on incorporating the multiple spatial scales and temporal complexities (e.g., simultaneity and disjuncture) involved in individual hybrid-accessibility. The following example, derived from the case examined in Kwan (2000d), illustrates the procedures for constructing the multiscale 3D extensibility diagram and its use in a GIS-based 3D geovisualization environment. The first step is to determine the most appropriate spatial scales and extract the relevant base maps from various digital sources.

Consider a person who lives and works in Franklin County, Ohio, and engages in cyberspatial activities (e.g., sending and receiving e-mail) involving cities in the northeastern region of the United States and other countries (e.g., South Africa and Japan). To prepare the GIS base maps at these three spatial scales (local, regional, and global), a map of Franklin County and a regional map of 15 U.S. states in the northeastern part of the country were first extracted from commercial geographic database. Franklin County is the home county of the person in question, whereas the U.S. region extracted will be used to locate the three American cities involved in her cyber-transactions. These cities are Chicago in Illinois, Maywood in New Jersey, and Charlotte in North Carolina. At the global scale, the world map layer was derived from the digital map data that came with ArcGIS (many high-latitude regions and islands in the world map layer were eliminated for improving visual clarity).

After performing map-scale transformations to register these three map layers to the person's home location in Franklin County, these 2D map layers were converted to 3D shape files and added to an ArcGIS 3D Analyst scene as 3D themes. After preparing these map layers, 3D shape files for the person's space-time paths were generated using Avenue scripts and added to the 3D scene. These procedures finally created the multiscale extensibility diagram shown in Figure 6.4. It shows how various types of transactions at different spatial scales can be represented in a 3D GIS environment.

Figure 6.5 shows five types of activities undertaken on a particular day. On this day, the person in question worked from 8:30 a.m. to 5:30 p.m., and had a one-hour lunch break at a nearby restaurant (*c* on the diagram). She subscribes to a Web-casting service where news is continuously forwarded to her Web browser. On this day, she read some news about Yugoslavia, South Africa, and Nashville, Tennessee (*a* on

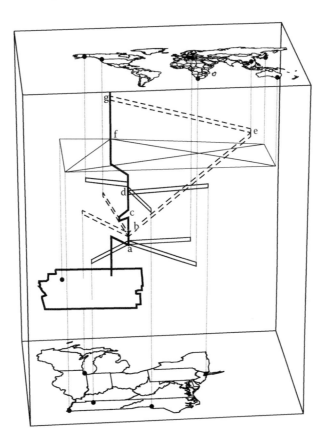

FIGURE 6.4   An extensibility diagram rendered using 3D GIS. (Source: Figure 14.3 in Kwan, M.-P., Human extensibility and individual hybrid-accessibility in space-time: A multiscale representation using GIS. In Information, Place, and Cyberspace: Issues in Accessibility, ed. D. G. Janelle, and D. C. Hodge, 241–56, © 2000 by Springer-Verlag. With kind permission of Springer Science and Business Media.)

the diagram) before she started work. An hour later she sent an e-mail message to three friends located in Hong Kong, Chicago, and Vancouver (*b*). The friend in Chicago read the e-mail two hours later and the friend in Vancouver read the e-mail five hours later. The friend in Hong Kong read the e-mail 13 hours later and replied immediately (*e*). The reply message from this friend, however, was read at 2:00 a.m. at the person's home (*g*). In the afternoon, she browsed Web pages hosted in New York, Charlotte, and Anchorage, Alaska (*d*). She was off from work at 5:30 p.m. and spent the evening at home. At 9:00 p.m. she started an ICQ (real-time chat) session with friends in Tokyo; Melbourne, Australia; Memphis, Tennessee, and Dublin, Ohio (*f* on the diagram).

As shown in Figure 6.4, very complex interaction patterns in cyberspace can be represented using multiple and branching space-time paths. These include temporally coincidental (real-time chat) and temporally noncoincidental (e-mailing) interactions; one-way radial (Web browsing), two-way dyadic or radial (e-mailing),

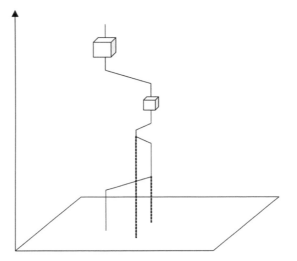

FIGURE 6.5   3D space-time path and information cube (© 2007 by Kwan, M.-P. and F. Ren).

and multiway (chat) interactions; incoming (Web casting) and outgoing (e-mailing) transactions. The method is thus capable of capturing the spatial, temporal, and morphological complexities of a person's extensibility in cyberspace.

Further, the visualization functions available in ArcGIS 3D Analyst also enable one to interactively explore the 3D scene in a very flexible manner (e.g., the scene is visible in real-time while zooming in and out, or rotating). This allows for the selection of the best viewing angle and is a very helpful feature especially when visualizing very complex space-time paths. To focus on only one type of transaction or activity at a particular spatial scale, one can select the relevant themes for display while keeping the other themes turned off. Further, when the three sets of paths and base maps are displayed at the same time, they can be color-coded to facilitate the visualization. In the original color 3D scene, each segment of the space-time paths are represented using the same color as the relevant base map (e.g., blue for Franklin County and local activities), conveying a rather clear picture of the spatiality and temporal rhythm characterizing the person's activities on that day. But in the black-and-white version presented in Figure 6.4, spike lines are used to identify the location involved in each transaction.

VISUALIZING CYBERSPATIAL ACCESSIBILITY USING INFORMATION CUBES

Although the extensibility diagram is helpful in revealing an individual's physical and cyberspatial activities, it is difficult to show any aggregate patterns in HSTB using this method. In one of our ongoing projects, a method for representing the number of cyberspatial opportunities a person can access while taking into account the person's daily space-time path was developed (Ren and Kwan 2006). The study used an activity-Internet diary data collected in Columbus, Ohio, in 2003 to 2004. The premise for this project is that people's access to the Internet is also restricted by the space-time constraints they face in their everyday life. For instance, many people can access the Internet only at certain places (e.g., workplace or home) or at certain times. The number of cyberspatial opportunities they can access can be analyzed with their space-time paths.

In the study we use the number of Web sites visited as a proxy for the number of cyber-opportunities to which a person has access. We made the simplifying assumption that all Web sites offer the same number of cyberspatial opportunities, and each Web site is therefore equally weighed. The more Web sites a person visited in a given amount of time, the greater the person's cyberspatial accessibility. Using data on the duration of Internet usage, the mode of Internet connection (e.g., broadband), and the number of Web sites visited, information cubes were constructed along an individual's space-time line paths as shown in Figure 6.5. Each information cube represented the volume of the information space a person has access to at a particular time and location. The volume of all information cubes along a person's daily space-time path can be used as a measure of that person's cyber-accessibility on that particular day.

With this modified representation of space-time paths, we will be able to investigate Internet activity patterns of different social groups. First, aggregating information cubes along the time line will allow us to compare how information spaces of particular social groups change in a day. Second, it is also feasible to aggregate information cubes based on activity purposes. In doing so, we will gain new insights into what people do via the Internet and what their cyberspatial accessibility is for different activity purposes (e.g., e-banking). We can then examine whether any differences exist among social groups.

Figure 6.6 shows the toolbar with customized function for visualizing 3D space-time paths and the 3D space-time path of a selected woman. The button "LoadData" accesses the geodatabase and loads all persons' identifier for future selection. With any selected person, the system will ask for the day and then display the space-time path for the required day for the selected person. In addition, the system also generates aggregate space-time paths for different subgroups, including men and women working full time, part-time employed women, and nonemployed women.

As shown in Figure 6.6, the 3D space-time path and information cubes with different colors convey the information about where, when, and for what purpose an activity was undertaken. Activity purposes in this study are classified into five categories: work and work-related activities, household needs, personal needs, recreational activities, and social activities. By viewing the 3D space-time path of the selected woman, we will have an idea about her spatial trajectory in a day, including working activity at office, having lunch at Heavenly Ham, working at a Microsoft store, running errands at home, Staples, and PetSmart. There were four Internet activities: two were for work purpose (in red) in the morning and afternoon, one was conducted for personal needs (in brown) at the office at noon, and the last one was performed at home for household needs (in green) around 6 p.m. Using this method, a person's cyberspatial activities can be visualized together with her/his space-time path.

Geocomputational Methods

The term geocomputation refers to an array of activities involving the use of new computational tools and methods to depict geographical variations of phenomena

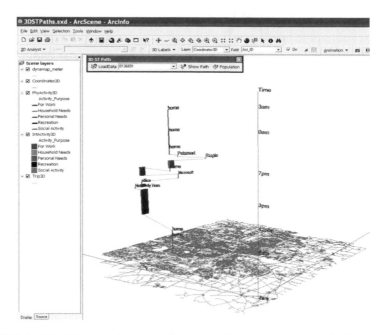

FIGURE 6.6   The 3D space-time path of a selected woman in the sample (From Ren and Kwan © 2007).

across scales (Longley 1998). It encompasses a wide range of computer-based techniques, including expert systems, fuzzy sets, genetic algorithms, cellular automata, neural networks, fractal modeling, visualization, and data mining. Many of these methods are derived from the field of artificial intelligence and the more recently defined area of computational intelligence (Couclelis 1998). The availability of affordable high-speed computing and the development of GIS technologies in recent years have greatly facilitated the application of geocomputation in time-geographic research.

The use of geocomputation as a time-geographic method is most visible in recent research on individual accessibility (Kim and Kwan 2003; Kwan 1998, 1999b; Miller 1999; O'Sullivan et al. 2000; Weber 2003; Weber and Kwan 2002, 2003). It involves the development and application of dedicated algorithms for computing space-time accessibility measures within a GIS environment. Space-time accessibility measures are largely based on the analytical framework formulated by Lenntorp (1976) and Burns (1979). They are based on the time-geographic construct of potential path area, which is the geographic area that can be reached within the space-time constraints established by an individual's fixed activities. It is the area that an individual can physically reach after one fixed activity ends while still arriving in time for the next fixed activity. All space-time accessibility measures are derived from certain measurable attributes of this area (e.g., number of opportunities it includes).

Because of the need to represent real-world complexities and to deal with the large amount of geographic data, GIS provides an effective environment for implementing geocomputational algorithms for space-time accessibility measures. With modern

GIS technologies and increasingly available disaggregate data, highly refined space-time measures of individual accessibility can be made operational. Several studies in recent years have developed and implemented geocomputational algorithms based on the time-geographic perspective. Drawing on the author's recent search, several examples are discussed below to illustrate the application of geocomputation in time-geographic research.

COLUMBUS STUDY

The first major effort in the geocomputation of space-time accessibility measures is that by Kwan (1998). The study examined individual access to urban opportunities for a sample of 39 men and 48 women in Columbus, Ohio. Data for the study came from three main sources. The first source is an activity-travel diary dataset collected by the author through a mail survey in 1995. In addition to questions about the activity-travel characteristics of the respondent, data of the street addresses of all activity locations and the subjective spatial and temporal fixity ratings of all out-of-home activities were collected (Kwan 2000a). The second source of data is a digital geographic database of the study area that provides detailed information about all land parcels, their attributes, and other geographical features of the study area. Among the 34,442 nonresidential parcels in the database, 10,727 parcels belonging to seven land-use categories were selected as the urban opportunities in the study. The third data source is a detailed digital street network of the study area. The network database contains 47,194 arcs and 36,343 nodes of Columbus streets and comes with comprehensive address ranges for geocoding locations.

Using these data, 20 conventional measures of the gravity and cumulative opportunity variants were evaluated using the home locations of the 87 individuals as origins and 10,727 property parcels as destinations. Distances were computed using point-to-point travel times through a digital street network. Three space-time measures were also computed for each individual using a geocomputational algorithm written in ARC Macro Language (AML) and implemented in ARC/INFO GIS. These three measures evaluate the size of the space that can be reached, the number of opportunities that can be reached, and the size or attractiveness of those opportunities. The algorithm used in the study was based on the one developed in Kwan and Hong (1998). Although it only provides an approximate solution of the exhaustive set of reachable opportunities, it is computationally more tractable. It uses the intersection of a series of paired arc-allocations to generate individual network-based PPAs (potential path areas), each of which is defined by the space-time coordinates of two fixed activities (see Figure 6.7 for a schematic representation of the algorithm).

The results of the study reveal the contrast between conventional and space-time measures. While the values produced by most gravity and cumulative opportunity measures were highly correlated and produced similar spatial patterns, space-time measures were very different. Gravity measures tended to replicate the geographical patterns of urban opportunities in the study area by favoring areas near major freeway interchanges and commercial developments, while cumulative opportunity measures emphasized centrality within the city by showing the downtown area to be the most accessible place. In contrast, space-time measures produced different

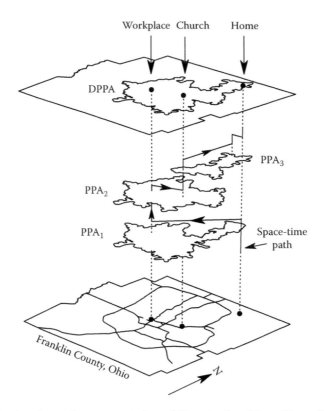

FIGURE 6.7   A schematic representation of Kwan's algorithm. (From Kwan, M.-P., Gender and individual access to urban opportunities: A study using space-time measures, The Professional Geographer 51: 210–27, Blackwell, 1999. Use by permission of Blackwell Publishing.)

spatial patterns, and the patterns for men resembled the spatial distribution of opportunities in the study area while the women's patterns were considerably different. The study shows that space-time measures are capable of revealing individual differences that are invisible when using conventional accessibility measures.

PORTLAND STUDY 1

To take into account the effect of the spatial and temporal variations in travel speeds and facility opening hours on individual accessibility, Weber and Kwan (2002, 2003) developed a second-generation algorithm for computing space-time accessibility measures. The study used a new geographic database with the enhanced geocomputational algorithm. The activity-travel diary data set used was collected through the Activity and Travel Survey in the Portland Metropolitan Area in Oregon in 1994 and 1995. The dataset logged a total of 129,188 activities and 71,808 trips undertaken by 10,084 respondents. Among the respondents, 101 men and 99 women were selected for the study. Besides this, a digital street network with estimates of free flow and congested travel times (with 130,141 arcs and 104,048 nodes) and a comprehensive geographic database of the study area were used. A digital geographic database containing 27,749 commercial and industrial land parcels was used to represent potential activity opportunities in the study area.

The analytical procedures involved creating a realistic representation of the temporal attributes of the transport network and urban opportunities in the study area, as well as developing a geocomputational algorithm for implementing space-time accessibility measures within a GIS environment. The algorithm was developed and implemented using Avenue, the object-oriented scripting language in the ArcView 3.x GIS environment. Five space-time accessibility measures were computed. The first is the length of the road segments contained within the daily potential path area (DPPA). The second is the number of opportunities within the DPPA. The total area and total weighted area of the land parcels within the DPPA is the third and fourth space-time accessibility measures computed. Finally, to incorporate the effect of business hours on accessibility measures, opportunity parcels were assumed to be available (and could therefore be accessible to an individual) only from 9:00 a.m. to 6:00 p.m. This creates the fifth accessibility measure.

The results show that link-specific travel times produce very uneven accessibility patterns, with access to services and employment varying considerably within the study area. The time of day activities were carried out has also been shown to have an effect on accessibility, as evening congestion sharply reduced individuals' access throughout the city. The effect of this congestion on mobility is highly spatially uneven. Further, the use of business hours to limit access to opportunities at certain times of the day shows that nontemporally restricted accessibility measures produce inflated values by treating these opportunities as being available at all times of the day. It is not just that incorporating time reduces accessibility, but that it also produces a very different, and perhaps unexpected, geography of accessibility (Weber and Kwan 2002). This geography depends much on individual behavior and so cannot be discerned from the location of opportunities or congestion alone. The study observed that the role of distance in predicting accessibility variations within cities is quite limited (Kwan and Weber 2003).

PORTLAND STUDY 2

In an attempt to render earlier geocomputational algorithms more realistic, several enhancements were conceived and implemented by the third-generation algorithm developed by Kim and Kwan (2003). First, space-time accessibility is extended as a measure of not only the number of accessible opportunities, but also the duration for which these facilities can be enjoyed given the space-time constraint of an individual and facility opening hours. Second, more realistic travel times are incorporated through better representation of the transportation network, such as one-way streets in downtown areas and turn prohibition — besides incorporating the effect of congestion and location- and segment-specific travel speeds. Third, ways are developed to better incorporate other factors such as facility opening hours, minimum activity participation time, maximum travel-time threshold, and delay times. The study seeks to enhance space-time accessibility measures with more rigorous representation of the temporal and spatial characteristics of opportunities and human activity-travel behavior.

A new GIS-based geocomputational algorithm was developed to implement these enhancements. The key idea of the algorithm is to efficiently identify all of the feasible opportunities within the space-time prism using several spatial search operations in ArcView GIS, while limiting the spatial search boundary with information about the travel and activity participation time available between two fixed activities. This algorithm was developed based on numerous tests of the computational efficiency of different methods and a series of experiments using a large activity-travel diary data set and a digital street network. The GIS algorithm for deriving the PPA and for calculating space-time accessibility was implemented using Avenue in ArcView GIS. The study shows that space-time accessibility measures that do not consider the effect of facility opening hours and activity duration threshold will tend to over-estimate individual accessibility.

CONCLUSION

This chapter suggests that the time geographic perspective provides a sound conceptual framework for developing methods for analyzing space-time paths and HSTB. We discussed two types of the GIS-based time geographic methods developed in recent years: three-dimensional (3D) geovisualization and geocomputational methods. Usefulness of these methods is illustrated through examples drawn from recent studies by us and other researchers. Besides these two types of methods, there are also many methodological innovations in the study of HSTB in recent years. There are studies that use an object-oriented approach for extracting and visualizing space-time paths. There are studies that attempt to reconstruct space-time paths from large geographic database using neural networks and self-organizing maps. Researchers have also used agent-based modeling and pattern aggregation techniques, such as sequence alignment to extract meaningful patterns of HSTB. These are promising areas for future research in GIS-based time-geographic methods.

In terms of substantive foci, interest is increasing in extending time-geographic research to the study of how information and communications technologies (ICT) influence human activity-travel patterns in recent years. These studies seek to describe and analyze the impact of Internet and mobile phone use on people's behavior — for example, will e-shopping or e-banking reduce people's trips to shops or banks in the physical world? These studies also try to understand how ICT use may affect people's space-time constraints — for example, will people undertake more social and recreational activities by using the time the saved through using ICT (like e-shopping)? Both the substantive and methodological development in analytical methods for representing life paths and understanding HSTB in the urban context will further enhance our knowledge in this important area of research.

REFERENCES

Adams, P. C. 1995. A reconsideration of personal boundaries in space-time. *Annals of the Association of American Geographers* 85:267–85.

Battelle. 1997. *Lexington Area Travel Data Collection Test: Final Report*. Battelle Memorial Institute, Columbus, OH.

Bhat, C. R. and S. K. Singh. 2000. A comprehensive daily activity-travel generation model system for workers. *Transportation Research* A 34(1):1–22.

Burns, L. D. 1979. *Transportation, Temporal, and Spatial Components of Accessibility*. Lexington, MA: Lexington Books.

Carlstein, T., D. Parkes, and N. Thrift. 1978. *Timing Space and Spacing Time II: Human Activity and Time Geography*. London: Arnold.

Chapin, F. S., Jr. 1974. *Human Activity Patterns in the City*. New York: John Wiley and Sons.

Cullen, I., V. Godson, and S. Major. 1972. The structure of activity patterns. *In Patterns and Processes in Urban and Regional Systems*, ed. A. G. Wilson, 281–296. London: Pion.

Couclelis, H. 1998. Geocomputation in context. In Geocomputation: A Primer, ed. P. Longley, S. M. Brooks, R. McDonnell, and B. MacMillan. New York: John Wiley & Sons.

Forer, P. 1998. Geometric approaches to the nexus of time, space, and microprocess: Implementing a practical model for mundane socio-spatial systems. *In Spatial and Temporal Reasoning in Geographic Information Systems*, ed. M. J. Egenhofer and R. G. Golledge. Oxford: Oxford University Press.

Frihida, A., D. J. Marceau, and M. Thériault. 2004. Development of a temporal extension to query travel behavior time paths using an object-oriented GIS. GeoInformatica 8(3):211–35.

Golob, T. F. 1985. Analyzing activity pattern data using qualitative multivariate statistical methods. *In Measuring the Unmeasurable*, ed. P. Nijkamp, H. Leitner, and N. Wrigley, 339–356. Boston: Martinus Nijhoff.

Golob, T. F. and M. G. McNally. 1997. A model of activity participation and travel interactions between household heads. *Transportation Research* B 31(3):177–194.

Goulias, K.G. 1999. Longitudinal analysis of activity and travel pattern dynamics using generalized mixed Markov latent class models. *Transportation Research* B 33(8):535–558.

Hägerstrand, T. 1970. What about people in regional science? *Papers of Regional Science Association* 24:7–21.

Hendricks, M. D., M. J. Egenhofer, and K. Hornsby. 2003. Structuring a wayfinder's dynamic space-time environment. *COSIT 2003*:75–92.

Janelle, D. G. 1973. Measuring human extensibility in a shrinking world. *Journal of Geography* 72:8–15.

Kim, H.-M. and M.-P. Kwan. 2003. Space-time accessibility measures: A geocomputational algorithm with a focus on the feasible opportunity set and possible activity duration. *Journal of Geographical Systems* 5:71–91.

Kwan, M.-P. 1998. Space-time and integral measures of individual accessibility: A comparative analysis using a point-based framework. *Geographical Analysis* 30:191–216.

Kwan, M.-P. 1999a. Gender, the home-work link, and space-time patterns of non-employment activities. *Economic Geography* 75:370–94.

Kwan, M.-P. 1999b. Gender and individual access to urban opportunities: A study using space-time measures. *The Professional Geographer* 51:210–27.

Kwan, M.-P. 2000a. Gender differences in space-time constraints. *Area* 32:145–56.

Kwan, M.-P. 2000b. Interactive geovisualization of activity-travel patterns using three-dimensional geographical information systems: A methodological exploration with a large data set. *Transportation Research* C 8:185–203.

Kwan, M.-P. 2000c. Evaluating Gender Differences in Individual Accessibility: A Study Using Trip Data Collected by the Global Positioning System. *A Report to the Federal Highway Administration* (FHWA), U.S. Department of Transportation, 400 Seventh Street, S.W., Washington, D.C. 20590.

Kwan, M.-P. 2000d. Human extensibility and individual hybrid-accessibility in space-time: A multi-scale representation using GIS. In *Information, Place, and Cyberspace: Issues in Accessibility*, ed. D. G. Janelle, and D. C. Hodge, 241–56. Berlin: Springer-Verlag.

Kwan, M.-P. 2001a. Analysis of LBS-derived data using GIS-based 3D geovisualization. Paper presented at the Specialist Meeting on Location-Based Services, Center for Spatially Integrated Social Science (CSISS), University of California, Santa Barbara, December 14–15, 2001.

Kwan, M.-P. 2001b. Cyberspatial cognition and individual access to information: the behavioral foundation of cybergeography. *Environment and Planning B* 28:21–37.

Kwan, M.-P. 2004. GIS methods in time-geographic research: Geocomputation and geovisualization of human activity patterns. *Geografiska Annaler* B 86(4):267–280.

Kwan, M.-P. and X.-D. Hong. 1998. Network-based constraints-oriented choice set formation using GIS. *Geographical Systems* 5:139–62.

Kwan, M.-P. and J. Lee. 2004. Geovisualization of human activity patterns using 3D GIS: A time-geographic approach. In *Spatially Integrated Social Science*, ed. M. F. Goodchild and D. G. Janelle, 48–66. New York: Oxford University Press.

Kwan, M.-P. and J. Weber. 2003. Individual accessibility revisited: Implications for geographical analysis in the twenty-first century. *Geographical Analysis* 35:341–53.

Laws, G. 1997. Women's life courses, spatial mobility, and state policies. In *Thresholds in Feminist Geography: Difference, Methodology, Representation*, ed. J. P. Jones III, H. J. Nast, and S. M. Roberts, 47–64. New York: Rowman and Littlefield.

Lenntorp, B. 1976. Paths in Time-Space Environments: A Time Geographic Study of Movement Possibilities of Individuals. *Lund Studies in Geography B: Human Geography*, Gleerup, Lund.

Lenntorp, B. 1978. A time-geographic simulation model of individual activity programs. In *Timing Space and Spacing Time Volume 2: Human Activity and Time Geography*, ed. T. Carlstein, D. Parkes, and N. Thrift. London: Edward Arnold.

Longley, P. 1998. Foundations. In *Geocomputation: A Primer*, ed. P. Longley, S. M. Brooks, R. McDonnell, and B. MacMilllan. New York: John Wiley & Sons.

Lowe, J. W. 2003. Special handling of spatio-temporal data. *Geospatial Solutions* 13(11):42–45.

Ma, J. and K. G. Goulias. 1997a. An analysis of activity and travel patterns in the Puget Sound transportation panel. In *Activity-based Approaches To Travel Analysis*, ed. D. Ettema, and H. Timmermans, H., 189–207. Tarrytown, NY: Elsevier Science.

Ma, J. and K. G. Goulias. 1997b. A dynamic analysis of person and household activity and travel patterns using data from the first two waves in the Puget Sound Transportation Panel. *Transportation* 24(3):309–331.

Mårtensson, S. 1977. Childhood interaction and temporal organization. *Economic Geography* 53:99–125.

McCormack, E. 1999. Using a GIS to enhance the value of travel diaries. *ITE Journal* 69(1):38–43.

Miller, H. J. 1991. Modelling accessibility using space-time prism concepts within geographic information systems. *International Journal of Geographical Information Systems* 5:287–301.

Miller, H. J. 1999. Measuring space-time accessibility benefits within transportation networks: basic theory and computational procedures. *Geographical Analysis* 31:187–212.

Miller, H. J. and Y.-H. Wu. 2000. GIS software for measuring space-time accessibility in transportation planning and analysis. *GeoInformatica* 4:141–59.

Murakami, E. and D. Wagner. 1999. Can using global positioning system (GPS) improve trip reporting? *Transportation Research* C 7:149–65.

Odland, J. 1998. Longitudinal analysis of migration and mobility spatial behavior in explicitly temporal contexts. In *Spatial and Temporal Reasoning in Geographic Information Systems*, ed. M. J. Egenhofer and R. G. Golledge. Oxford: Oxford University Press.

Ohmori, N., Y. Muromachi, N. Harata, and K. Ohto. 1999. A study on accessibility and going-out behavior of aged people considering daily activity pattern. *Journal of the Eastern Asia Society for Transportation Studies* 3:139–153.

O'Sullivan, D., A. Morrison, and J. Shearer. 2000. Using desktop GIS for the investigation of accessibility by public transport: An isochrone approach. *International Journal of Geographical Information Science* 14:85–104.

Parkes, D. N. and N. Thrift. 1975. Timing space and spacing time. *Environment and Planning* A 7:651–70.

Recker, W. W., M. G. McNally, and G. S. Root. 1987. An empirical analysis of urban activity patterns. *Geographical Analysis* 19(2):166–181.

Ren, F. and M.-P. Kwan. 2007. Geovisualization of human hybrid activity-travel patterns. *Transactions in GIS* 11(5):721–44.

Rollinson, P. 1998. The everyday geography of the homeless in Kansas City. *Geografiska Annaler* B 80:101–15.

Shoval, N. and M. Isaacson. 2007. Sequence alignment as a method for human activity analysis in space and time. *Annals of the Association of American Geographers* 97(2):281–96.

Takeda, Y. 1998. Space-time prisms of nursery school users and location-allocation modeling. *Geographical Sciences* (Chiri-kagaku, in Japanese) 53:206–216.

Thrift, N. 1977. An Introduction to Time Geography. Geo Abstracts, University of East Anglia, Norwich.

Tivers, J. 1985. *Women Attached: The Daily Lives of Women with Young Children*. London: Croom Helm.

Weber, J. 2003. Individual accessibility and distance from major employment centers: An examination using space-time measures. *Journal of Geographical Systems* 5:51–70.

Weber, J. and M.-P. Kwan. 2002. Bringing time back in: a study on the influence of travel time variations and facility opening hours on individual accessibility. *The Professional Geographer* 54:226–40.

Weber, J. and M.-P. Kwan. 2003. Evaluating the effects of geographic contexts on individual accessibility: A multilevel approach. *Urban Geography* 24(8):647–71.

Yu, H. and S.-L. Shaw. 2004. Representing and visualizing travel diary data: A spatio-temporal GIS approach. *Proceedings of the 2004 ESRI International User Conference*, San Diego, CA.

7 Relating Visual Changes in Images with Spatial Metrics

Nina Lam, Guiyun Zhou, and Wenxue Ju

CONTENTS

INTRODUCTION

Visualizing land cover and feature changes from time-series remote sensing images is the first and fundamental step in understanding the change dynamics of many geographic phenomena. In general, changes that are drastic and spatially concentrated can be discerned more easily, whereas small and gradual changes that scatter are very difficult to detect. Moreover, human visualization is not sufficient to detect changes continuously and rapidly. We need computation and measurement to support visualization and help identify where and what are the visual changes. Visual representation alone cannot satisfy analytical needs (Thomas and Cook 2005). Therefore, relating the change dynamics observed visually with quantitative measures that can be computed from the images is very much needed to enhance the visualization capability. These measures or metrics would enable automation and hence rapid and continuous monitoring, which is crucial to supporting timely decision-making and risk assessment especially during extreme events (e.g., terrorist attacks, hurricanes, forest fires, earthquakes, disease spread).

At the same time, we need better visualization methods to display the results from measurements so that the most important information can be visualized quickly and easily. This necessitates two considerations. First, when analyzing land use and land cover changes derived from time-series remote sensing images, some changes may not be real, as they may be due to difference in cloud cover, misregistration of images, seasonal change, and other extraneous effects. The remote sensing community has widely recognized this problem and has developed various approaches to minimize these effects. We argue in this chapter that the use of spatial metrics on band-ratio images for change detection would help in reducing the extraneous effects and detecting real changes. The main notion is that the use of spatial metrics, in combination with original spectral information in band-ratio form, will increase land cover and feature classification accuracy, as well as have great potential for rapid change detection (Lam, forthcoming). Second, the visualization routine should be able to highlight the largest or most important changes automatically, so that it can be used as guidance for interpreters to focus on areas that need attention most. Such seemingly simple routines are not easily applied in commercial image processing packages. Specialized computer software will need to be developed to facilitate both measurement and visualization concurrently.

Although a huge literature on change detection using remotely sensed images exists, and new approaches continuously appear in the literature, we are still far from being able to automate the change detection process. The high variability of ground conditions as manifested in individual and time-series imagery, as well as many other sources of uncertainty including the scale issues (Lam et al. 2004; McMaster and Usery 2004), make the process very difficult to generalize and automate. The search for useful approaches and methods for characterizing and modeling landscape dynamics, especially rapid land cover identification and change detection, remains a major research challenge in these fields (Bian and Walsh 2002).

The purpose of this chapter is to introduce the use of spatial metrics as an approach to detecting the change dynamics from time-series images and to link visualization and measurement in a seamless manner. The ultimate goal is to automate the change-detection process through close coupling of measurement and visualization. This chapter will first discuss the criteria for selecting appropriate spatial metrics for rapid change detection, since there are lots of spatial and textural indices in various literature, and not all of them are applicable. We will then introduce the use of classical Moran's I spatial autocorrelation statistic, a spatial index that meets all the criteria, in change detection. Two examples illustrating the approach to linking visual changes with spatial metrics are demonstrated, one using digital camera images and the other using Landsat-TM images of New Orleans. The prospect and problems of using spatial metrics for change dynamic analysis are then evaluated.

CRITERIA FOR SELECTING SPATIAL METRICS

These are numerous spatial or textural measures in the literature. Lam (forthcoming) outlined some of the commonly used spatial metrics, as well as the criteria for selecting a spatial metric for rapid change detection. For example, some measures are founded on vigorous statistical theory and mathematical derivation (e.g., fractals, wavelets, spatial autocorrelation), while others are based on geometric measurements with unknown statistical properties and theoretical minimum or maximum (e.g., edge density) (Haralick 1979; Haralick et al. 1973; Lam et al. 1998). Some metrics measure one aspect of the landscape (e.g., landscape composition), while others quantify another (e.g., landscape configuration) (McGarigal and Marks 1995). There are also attempts to classify various spatial measures. Baskent and Jordan (1995) classified landscape indices as areal, linear, or topological. Yet another classification of spatial and textural measures is based on whether the measures are applied directly to the original matrix (first-order textural measures), or to matrices derived from the original matrix such as the gray-level co-occurrence matrix or wavelet decomposed images (second-order textural measures) (Jensen 2004). Hence, it is imperative to develop useful criteria to evaluate or classify these textural measures.

What is a good spatial or textural measure? Although there may be no definite answers until each measure is tested extensively for their discriminating and explanatory power, the following criteria will be used to evaluate and guide our understanding of these various measures (Lam, forthcoming). Ideally, a good spatial measure should have the following properties:

1. *The spatial measure should be conceptually simple and easy to calculate and visualize.* For example, statistical mean and standard deviation are concepts easily grasped by most researchers and their statistical properties are well known. Extension of these basic statistical measures in a spatial domain with some modifications may provide a useful first approximation towards an understanding of the land cover/land use pattern being studied. On the contrary, some measures may be conceptually simple but require additional steps for calculation, such as edge density, which may need an additional step or algorithm to find the edges.

2. *The spatial measure should have theoretical maximum and minimum.* For example, Moran's I, a most commonly used spatial autocorrelation statistic, has a range of ± 1. A Moran's I value of 1.0 indicates a maximum positive spatial autocorrelation, a -1.0 indicates a maximum negative spatial autocorrelation, and a value of 0.0 indicates a random pattern (Cliff and Ord 1973).

3. *The spatial measure should reflect clearly and intuitively the characteristics of the image pattern in a consistent manner.* For example, a lower fractal dimension value means a less spatially complex image, therefore given an image computed with a fractal dimension value of 2.3, we should be able to infer that this image is far less complex than an image computed with a fractal dimension value of 2.7, and a visual display of the two images should be able to reveal the difference (Lam 1990). Fractal dimension (D) also has the second property, where D is expected to range from 2.0–3.0 for surfaces and 1.0–2.0 for lines (Mandelbrot 1982).

4. *The statistical properties of the spatial measure should be known to provide statistical confidence of the computed value.* For example, theoretically a Moran's I value of 0.0 indicates a random pattern. If a pattern yields a computed value of 0.2, can we determine if this value is statistically the same or different from 0.0 to conclude that the pattern is random or not? Fortunately, the statistical properties of Moran's I are relatively well known and hypothesis testing of whether a computed I value is significant can be conducted. Under the assumption of randomization, the first and second

moments of the Moran's *I* value can be computed and the statistical significance of the value determined (Goodchild 1986). On the contrary, the statistical properties of fractal dimension are still not clear, even though it has well-defined theoretical minimum and maximum. Hence, it is difficult to judge, for example, if an image with a fractal dimension of 2.3 is significantly different from another image with a fractal dimension of 2.4. It is noted that the statistical properties of most spatial measures are very difficult to derive and therefore remain unclear; many researchers have resorted to the Monte Carlo approach to develop empirical probability functions for statistical hypothesis testing (e.g., Openshaw 1989).

5. *The spatial measure can be computed globally for the entire study area or locally for a local neighborhood.* For example, some landscape metrics are only applicable at either the patch, class, or landscape levels, instead of being applicable to all levels (McGarigal 2002), whereas Moran's *I* and fractals, as implemented in a software module called ICAMS (Image Characterization And Modeling System), can be applied both globally and locally to capture local change (Lam 2004; Emerson et al. 2005).

6. Finally, *the spatial measure should be applicable directly to both classified and unclassified images.* For example, the landscape metrics in FRAGSTATS were developed exclusively for categorical maps (O'Neill et al. 1988; McGarigal 2002), or in other words, classified images, though some of the metrics can be modified so that they can be applied to unclassified images. On the contrary, fractals, Moran's *I*, local variance (Woodcock and Strahler 1987), variogram, and lacunarity (all available in ICAMS) can be applied to both unclassified and classified images.

This last property is considered very important to automated land cover and land use classification and change detection for two reasons. First, if they can be applied directly to unclassified images, changes in the images can be detected prior to the tedious classification process. Only after the changes are determined to be significant, then it is necessary to identify or classify what the changes are. This is considered a more efficient approach, especially for continuous environmental monitoring. Second, since these spatial indices measure the spatial variations among pixels instead of comparing pixel by pixel, they are more likely to reflect dominant changes rather than spurious changes that might have resulted from using images taken in different time periods. If there are only small changes in land cover, it is expected that the spatial relationship will not alter and the spatial index values will remain the same. On the contrary, if there are significant land-cover changes, then it is expected that the spatial structure will be altered, and the spatial indices that are designed to measure the spatial structure should be able to capture these changes.

METHODS

SPATIAL AUTOCORRELATION STATISTIC — MORAN'S I

Spatial autocorrelation statistics have been commonly used to measure the degree of clustering, randomness, or fragmentation of a spatial pattern. The two most common spatial autocorrelation measures for interval-ratio data are Moran's I and Geary's C (Cliff and Ord 1973). Moran's I is generally preferred over Geary's C, because the values of the former are more intuitive (i.e., positive values for positive autocorrelation and vice versa). Moran's I was also found to be generally more robust (Goodchild 1986). Moran's I is calculated from the following formula:

$$I(d) = \frac{n \sum_i^n \sum_j^n w_{ij} z_i z_j}{w \sum_i^n z_i^2} \tag{7.1}$$

where w_{ij} is the weight at distance d so that $w_{ij} = 1$ if point j is within distance d from point i; otherwise, $w_{ij} = 0$; z's are deviations from the mean for variable y, and w is the sum of all the weights where $i \neq j$. Moran's I varies from +1.0 for perfect positive correlation (a clumped pattern) to −1.0 for perfect negative correlation (a checkerboard pattern). A Moran's I value of 0.0 would indicate a random pattern.

Moran's I and Geary's C were originally developed for measuring polygonal data (e.g., geographical regions) where the number of units or points measured (n) is often smaller. Only recently have these two measures been applied to raster data such as remotely sensed images, where the number of units being measured ($n \times n$) is much larger (Emerson et al. 1999; Lam et al. 2002). This could affect the utility of Moran's I in characterizing image complexity. In a benchmark study using simulated fractal surfaces, Lam et al. (2002) found that for surfaces that have low fractal dimension (low complexity), Moran's I was found to be ineffective and did not reflect accurately the complexity of the simulated surfaces. But for surfaces with high dimension (e.g., remotely sensed images), Moran's I performed well and adequately reflected the spatial complexity of the surfaces, even though the range of Moran's I values was small. However, a major advantage of Moran's I is that, in addition to meeting all the criteria listed above, it is very stable computation-wide and with less ambiguity, and it can be applied with a small window size (Myint et al. 2004). Moran's I and fractal dimension (D) have an inverse relationship, in which a spatial pattern with a high degree of fragmentation will have a low Moran's I but a high fractal dimension (Lam et al. 2002).

IMAGE CHARACTERIZATION AND MODELING SYSTEM (ICAMS)

Both Moran's I and Geary's C have already been implemented in ICAMS, which was developed in the 1990s by the author and her collaborators (Lam et al. 1998 and 2002; Quattrochi et al. 1997). Other major indices have also been implemented, including fractals, wavelets, lacunarity, and variogram analysis. With ICAMS, these

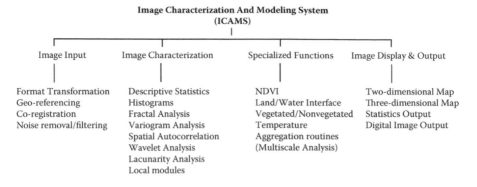

FIGURE 7.1   Main functions of the Image Characterization And Modeling System (ICAMS).

indices can be computed globally (applied to the entire study area) or locally by varying the moving-window size.

ICAMS has four subsystems: image input, image output, image characterization, and specialized functions (Figure 7.1). ICAMS was developed with an objective of providing scientists with innovative spatial analytical tools to visualize, measure, and characterize landscape patterns so that environmental conditions or processes can be assessed and monitored more effectively. ICAMS was originally designed in the '90s as a module compatible with Arc/Info and Intergraph-MGE platforms. The software has since been re-programmed as a standalone module that can run on Windows without any need for extra software. ICAMS can be obtained via the authors. This research will utilize ICAMS for the computation of Moran's *I*, change detection, and input/output display. It is envisioned that ICAMS will serve as a platform for incorporating results from future related research, adding more useful routines or generalized models for automated change detection, and improving interactive capabilities between visualization and computation.

VISUAL CHANGE DETECTION EXAMPLES

CAMERA IMAGES

Since the performance of using spatial indices for change detection is not well understood, we start with a set of simple images. Consider two photos taken from a digital camera at the backyard of a house in two time periods (Figure 7.2a and 7.2b). The first photo is without a person, and the second is with one, thus simulating a situation where there might be an intruder. The two images were taken with a time difference of less than one minute, and they were taken from a fixed photo stand, thus eliminating the possibility of errors from minor locational shifts. This controlled condition is rather unlikely for real satellite images, whereby the locations of two time-series images might not coincide exactly even after the geometric rectification process.

The RGB images (originally in JPEG format, then transformed to ERDAS/Imagine LAN format) were imported to ICAMS. Band 1 (red band) was selected for this experiment. The image size was 1536 × 2048 pixels. First, a difference image was created by subtracting the pre-image (without intruder) from the post-image (with

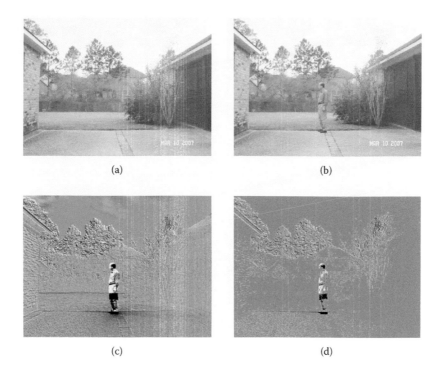

(a)   (b)

(c)   (d)

FIGURE 7.2 Digital camera images displayed using the red band (band 1); (a) without suspect; (b) with suspect; (c) difference image (with-without) displayed in a continuous mode; (d)
difference image displayed using 3 class intervals based on standard deviation unit of 2. The
brightest and darkest pixels in this image indicate the highest positive and negative changes
($>\pm2$ s.d.), whereas the gray pixels are pixels that have values within ±2 s.d.

intruder) directly without any transformation. Figure 7.2c displays the difference
image between the two time periods using a continuous mode (with a two-standard
deviation stretch). To enhance visualization of the largest changes, ICAMS can map
the difference image by dividing into classes using the standard deviation as class
intervals, however. The purpose of this type of mapping is to pinpoint the largest changes in both positive and negative directions. Figure 7.2d was mapped using
input parameters of 3 classes and standard deviation of 2. In other words, the first
interval, which has the darkest shade, contains the largest negative difference, with
pixel values below −2 standard deviations. The second class, which is in gray shade,
is for pixels that have difference values falling between ±2 standard deviations. The
third class, which is the brightest, is for pixels that have difference values >2 standard
deviations. Using this mapping method, the brightest and the darkest pixels in the
image are the pixels that have the largest positive and negative changes; hence logi-
cally attention should be paid to these areas first. This visualization method is simple
and flexible, and should guide the user in finding the most "important" changes
within a short amount of time. Another visualization option that is also available in
ICAMS is to map the pixels that have both the largest positive and negative changes

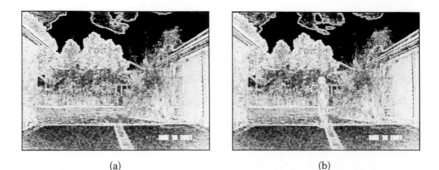

(a)   (b)

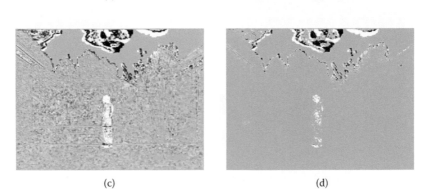

(c)   (d)

(e)   (f)

FIGURE 7.3 (a) and (b) Moran's *I*-transformed camera images using the red band and moving window of 17 × 17 pixels, without and with suspect, respectively; (c) and (d) difference image (with-without) of 17 × 17 window, displayed in continuous and discrete modes; (c) and (d) difference image of 9 × 9 window, displayed in continuous and discrete class modes.

using the brightest shade; this is used when the direction of changes (positive or negative) is assumed to be equal in terms of attention to be received.

The two images were transformed by calculating Moran's *I* using two moving-window sizes: 17 × 17 and 9 × 9 pixels. Difference images (Figure 7.3) were then created and displayed in both continuous and discrete modes using the same procedure as in Figure 7.2. The 17 × 17 window was chosen because a previous study by Kulkarni (2004), in which Landsat-TM images were used to study the impacts of

Hurricane Hugo along the South Carolina coast, found that this window size (corresponding to 510 m by 510 m) seemed to be the best in capturing the characteristics of natural land cover features. The 9×9 window size was arbitrarily chosen to demonstrate the effects of moving-window size on the change detection process.

An examination of the brightest and darkest pixels in the original difference image (Figure 7.2c) shows that the difference image clearly revealed the location of the intruder. Minor spectral differences appeared in other parts of the image, which were classified and shown as gray pixels (Figure 7.2d). The difference image shown here is relatively free of noise, largely because the time-series camera images were taken within a very short time span for a small area using a fixed stand. In most time-series satellite images, minor locational shifts as well as atmospheric effects in the images often occur, even after geometric rectification and atmospheric correction. This will result in more noises in the difference image than the current example, making it more difficult to detect changes by using the spectral difference alone. A preliminary analysis using photos that were taken without a fixed stand (not shown here) indicated that there were a lot of noises in the simple spectral difference image, and the intruder object was no longer as clear as this current difference image, because of other differences caused by minor locational shifting of the images. The minor misregistration between the two photos when taken without a fixed stand was found to contribute to the largest but unimportant changes, in addition to the real change where the intruder stood.

The Moran's I transformed images are difficult to interpret by themselves, which is expected, as they are not original images (Figure 7.3a and 7.3b). In general, the brighter the shade, the higher the Moran's I value, implying higher positive spatial autocorrelation and hence a smoother surface. On the contrary, the darker the shade, the lower is the spatial autocorrelation value (negative values), hence a more rugged surface. Edges and lines become more prominent and are easily observed in these texture-transformed images.

The transformed difference image (Figure 7.3c and 7.3d) shows that, as expected, the largest difference occurred in the intruder's location, as well as along the edges between the sky and the trees. However, large differences in Moran's I also occurred in the sky area, which is rather unexpected. An examination of the two transformed images found that the large differences occurred because some part of the sky area in the second image had very little variations but yielded high Moran's I values because of the relatively smooth surfaces within the window (e.g., for row = 65, column = 1507, $I = 0.7919$); whereas in the first image they were completely flat (same spectral reflectance value), leading to Moran's I values close to 0. In fact, the latter case resulted in a zero variance value that subsequently led to the computationally undefined problem. To avoid this computational problem, we replaced the zero variance with a very small value (0.00001), which then resulted in a Moran's I value close to 0. This unexpected result is inconsistent with the theoretical explanation of Moran's I, where a value of close to 0 indicates a random surface rather than a flat surface. As far as the authors know, this special condition (a flat surface) and inconsistency in Moran's I interpretation has seldom been reported in the literature, and a detailed analysis of the behavior of Moran's I is necessary but is beyond the scope of this paper.

(a)   (b)

(c)   (d)

FIGURE 7.4 (a) Original location of the intruder (T_2); (b) second location of the intruder (T_3); (c) and (d) difference image (T_3-T_2) displayed in continuous and discrete modes.

In terms of the effect of moving window size, when a smaller window size is applied to compute the texture, more details are revealed. The 9×9 transformed difference image (Figure 7.3e and 7.3f) shows the general trend as in the 17×17 transformed difference image but with more details. Based on visual comparison, it seems that for the current camera images, a 17×17 window is sufficient to reflect the changes. Hence it is used to demonstrate the next step.

To demonstrate the dynamics of changes, an additional photo depicting a change of the intruder location was taken to simulate a sequence of events. Figure 7.4a and 7.4b show the second (T_2) and the third (T_3) images. The spectral difference image between the third and the second images is shown in Figure 7.4c, and the Moran's I transformed difference image (using 17×17 windows) is shown in Figure 7.4d; both are displayed using the discrete class mode. Both images indicate clearly the location of the changes due to the movement of the intruder. This example shows that with additional images in a time sequence, more information can be revealed and the path of movement can be constructed.

These experiments using simple but real camera images show that misregistration and feature edges add difficulty in distinguishing real changes. The simple spectral difference image performs well because the camera images were taken from a fixed stand, thus eliminating most of the noises that were found in images taken without a fixed stand (images not shown). For textural transformed images such as Moran's I, the edge problem is more prominent. It is possible that in future research an edge filter can be developed and applied to remove these effects, so that the search for real changes can be more focused. This example also demonstrates the importance of visualization, especially how the algorithm can be made to determine what to visualize.

LANDSAT-TM IMAGES

We demonstrate in this example the use of spatial autocorrelation statistics for detecting and measuring changes from real satellite images. Two Landsat-TM images of New Orleans before and after Hurricane Katrina were used. Hurricane Katrina hit New Orleans on August 29, 2005. Two Landsat-TM images, dated November 7, 2004, and September 7, 2005, were obtained from the U.S. Geological Survey/ National Wetlands Research Center at Louisiana State University (LSU). Although not acquired exactly on the same anniversary dates, these two images were the best available for change detection. Both images have already been registered and geometrically rectified with a pixel resolution of 28.5 m prior to this study.

For this study, a subset of 512 × 512 pixels was created at the same location from each image. Each subset was then converted into a ratio image, in this case a normalized difference vegetation index (NDVI) image. The use of ratio instead of individual spectral bands will help minimize the errors that might arise from comparing time-series images, such as difference in cloud cover, atmospheric effects, sun angles, and/or topographic effects. NDVI is a ratio between infrared and red bands, and is computed by:

$$NDVI = (Infrared - Red)/(Infrared + Red) \qquad (7.2)$$

NDVI values range between ± 1.0. The index has been commonly used to delineate broad land cover classes such as vegetated versus nonvegetated areas, and water versus land. In general, clouds, snow, water, moist soil, and bright nonvegetated surfaces have NDVI values less than zero; rock and dry bare soils have values close to zero; and positive NDVI values generally indicate vegetated areas. In ICAMS, a value of -0.1 was set as the default threshold value to delineate water (< -0.1) and land (≥ -0.1). Similarly, a value of 0.25 was set as default value to delineate nonvegetated and vegetated boundaries. It is noted that such default values provide a starting point only, and the user will need to vary the threshold values to visualize and evaluate in order to find the values that best delineate the boundaries. Also, a fixed value may not work well in an urban setting like the present example.

Figure 7.5 displays the NDVI image subsets. The subsets are mainly confined to the Orleans Parish, with a small part of the Lower Ninth Ward in St. Bernard Parish shown at the east edge of the image (east of the Industrial Canal breach). The northwest corner of the image is Lake Pontchartrain, where significant storm surge has destroyed properties around the Lake. The three levee breaches at Industrial Canal, the 17th Street Canal, and the London Avenue Canal are also marked in the figure (Figure 7.5b). The NDVI difference image (post-pre) clearly shows the flooded areas of New Orleans, nine days after Katrina (Figure 7.5c). When displayed in discrete class mode (Figure 7.5d), largest negative changes in NDVI were found

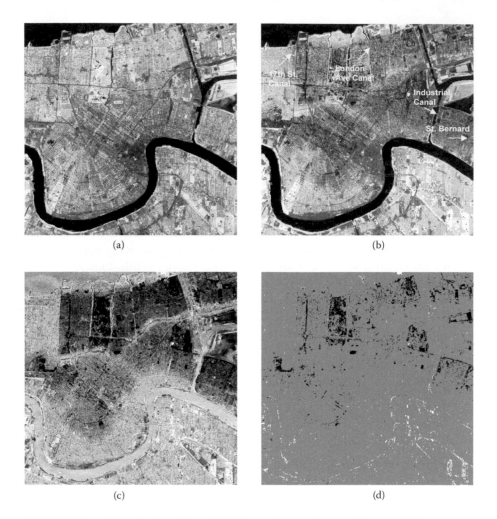

FIGURE 7.5 (a) Pre-Katrina (Nov. 7, 2004) and (b) post-Katrina (Sept. 7, 2005) Landsat-TM NDVI images; (c) and (d) difference image displayed in continuous and discrete modes.

in the north-central part of the image, including areas such as the City Park (close to the London Ave. Canal breach).

Using ICAMS, we computed the local Moran's I for each NDVI image using the same two window sizes, 9×9 and 17×17. Their difference images were then computed and displayed in both continuous and discrete modes (Figure 7.6). Table 7.1 lists the summary statistics of all the images. In general, the pre-Katrina NDVI image has more variation (coefficient of variation, CV = 0.34) than the post-Katrina NDVI image (CV = 0.23), due to a reduction of spectral variation because of the massive flooding. When the images are transformed, there is basically no difference in their coefficients of variation. Furthermore, when comparing the transformed images according to window size, larger moving-window size tends to create a smoother textural surface (higher mean I values) with higher nonspatial variation

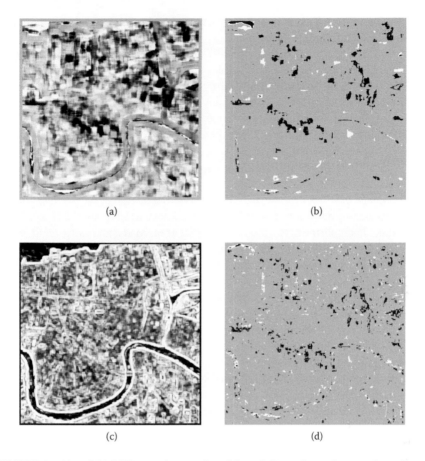

(a)

(b)

(c)

(d)

FIGURE 7.6 (a) and (b) Difference images from Moran's *I* transformed pre- and post-images using moving windows of 17 × 17 pixels, and displayed in continuous and discrete modes; (c) and (d) difference image of 9 × 9 pixels and displayed in continuous and discrete modes.

TABLE 7.1

Summary Statistics of Pixel Values in Pre- and Post-Katrina NDVI Images

	NDVI (scaled: 0-255)			Moran's *I* –Transformed (9 × 9)			Moran's *I* –Transformed (17 ×17)		
	Pre	**Post**	**Diff.**	**Pre**	**Post**	**Diff.**	**Pre**	**Post**	**Diff.**
Min	3.00	5.00	6.00	−0.24	−0.24	−0.68	−0.01	−0.03	−0.67
Max	255.00	255.00	255.00	0.94	0.94	0.71	0.97	0.97	0.66
Mean	130.21	148.15	148.68	0.54	0.53	0.01	0.64	0.64	−0.00
SD	43.84	33.82	20.98	0.17	0.17	0.12	0.15	0.15	0.09
CV	0.34	0.23	0.14	0.18	0.18	12.00	0.23	0.23	9.00

* SD – standard deviation; CV – coefficient of variation (= SD/mean); Difference image = (post-pre).

(higher CV values). This is in agreement with the observation made in Lam (1990) that high spatial variation in an image (as reflected by low Moran's I or high fractal dimension) is usually associated with low nonspatial variation (as reflected by standard deviation or coefficient of variation).

The Moran's I transformed NDVI difference images were first mapped using the continuous mode (with a two-standard deviation stretch) (Figure 7.5a and 7.5c). The darkest pixels represent the largest decrease in Moran's I value, implying an increase in spatial complexity. Similarly, the brightest pixels represent the largest increase in Moran's I value, implying a decrease in spatial complexity. As expected, the difference image derived from using the 9×9 moving-window has more details, whereas the difference image of 17×17 window shows the general trend. Based on the difference image of 9×9 window, we can observe that a large decrease in spatial complexity (increase in Moran's I — brightest pixels) occurred along the levee, the edge along Lake Pontchartrain, the Industrial Canal area, roads, and fairground, whereas large increase in complexity (darkest pixels) scattered in the Mid-City area.

The difference images were mapped using the discrete mode with parameters set as ±2 standard deviation and 3 classes (as in Figure 7.2). Using this method, the largest positive and negative changes in spatial complexity can be pinpointed. Both the difference images of 17×17 and 9×9 windows generated the same general locations of interest, with the former depicting the dominant trend and the latter showing smaller details. Again, based on the difference image of 9×9 window (Figure 7.6d), in general, large positive changes in Moran's I values (decrease in spatial complexity — brightest pixels) occurred at the northwest corner of the image (Lake Pontchartrain), 17th Canal Street breach, along the Mississippi River, and some spots south of the river. Large decrease of Moran's I value (increase in spatial complexity — darkest pixels) were found scattered throughout the Mid-City areas. The exact locations of these changes and their interpretations can be identified by overlaying other GIS layers such as the road maps, which is beyond the scope of this chapter.

In summary, this example shows that the textural approach can be useful in pinpointing the areas that need the most attention. With additional information layers, identification of the exact location of the changes and interpretation of why such changes occur can be made effectively. Additional real images in a time sequence will also be useful to improve the change detection ability and to increase our understanding of the land change dynamics after a major disaster. Other mapping methods (such as the regular density slicing method) could further be employed to enhance the mapping and visualization of these changes.

CONCLUSIONS

Visualization of dynamic changes needs analytic methods and measurement to quantify and help understand the observed changes and their underlying processes. At the same time, the need to visualize efficiently and intelligently will help in pushing the development of useful methods for change detection. Although change detection is an active research area in many fields with many algorithms and approaches developed, we are still far from being able to automate these tasks. The high variability of ground conditions as manifested in individual as well as time-series images makes it very difficult to generalize and automate. Despite the difficulty, automated methods that can narrow the search for changes that are significant and warrant attention are possible and will need to be developed. Relating visual changes with spatial metrics is an attempt toward this goal. We demonstrated in this paper how spatial metrics such as spatial autocorrelation statistics could be used in change detection, which is a first step towards the long-term goal of developing a suite of visual routines to support and facilitate rapid and reliable change detection for environmental monitoring and for further understanding of the change dynamics.

The results from the two examples show that using spatial metrics such as Moran's I computed directly from the images could add useful information. Although a special condition (flat surface) exists that may lead to spurious results, the approach of using Moran's I is simple and easy to apply. The areas of greatest changes identified from this approach will be best used by integrating information from other GIS data layers to help pinpoint and interpret where the changes are and what other environmental and social characteristics are associated with them. Algorithms that can remove the edge effects will also help in narrowing the search for meaningful changes. Future research is needed to experiment with more indices, explore how these indices are integrated with spectral data and other data layers, and develop algorithms to remove "meaningless" changes.

REFERENCES

Baskent, E.Z. and G.A. Jordan. 1955. Characterizing spatial structure of forest landscapes. *Canadian Journal of Forest Research* 25:1830–49.

Bian, L. and S.J. Walsh. 2002. Characterizing and modeling landscape dynamics: An introduction. *Photogrammetric Engineering and Remote Sensing* 68(10):999–1000.

Cliff, A.D. and J.K. Ord. 1973. *Spatial Autocorrelation*. New York: Methuen.

Emerson, C.W., N.S.N. Lam, and D.A. Quattrochi. 1999. Multiscale fractal analysis of image texture and pattern. *Photogrammetric Engineering and Remote Sensing* 65(1):51–61.

Emerson, C.W., N.S.N. Lam, and D.A. Quattrochi. 2005. A comparison of local variance, fractal dimension, and Moran's I as aids to multispectral image classification. *International Journal of Remote Sensing* 26(8):1575–88.

Goodchild, M.F. 1986. Spatial Autocorrelation. *CATMOG* (Concepts and Techniques in Modern Geography), no. 47. Norwich, England: Geo Books.

Gopal. S. and C. Woodcock. 1996. Remote sensing of forest change using artificial neural networks. *IEEE Transactions on Geoscience and Remote Sensing* 34(2):398–404.

Haralick, R.M. 1979. Statistical and structural approaches to texture. *Proceedings of IEEE* 67(5):786–804.

Haralick, R.M., K. Shanmugan, and J. Dinstein. 1973. Textural features for image classification. *IEEE Transaction on Systems, Man, and Cybernetics* 3(6):610–21.

Jensen, J. 2004. *Introductory Digital Image Processing: A Remote Sensing Perspective*, 3rd ed. Upper Saddle River, NJ: Prentice Hall.

Kulkarni, A. 2004. Evaluation of the Impacts of Hurricane Hugo on the Land Cover of Francis Marion National Forest, South Carolina Using Remote Sensing. M.S. thesis, Louisiana State University, Baton Rouge, Louisiana.

Lam, N.S.N. 1990. Description and measurement of Landsat TM images using fractals. *Photogrammetric Engineering and Remote Sensing* 56(2):187–95.

Lam, N.S.N. 2004. Fractals and scale in environmental assessment and monitoring. In *Scale and Geographic Inquiry: Nature, Society, and Method*, ed. E. Sheppard and R. McMaster, 23–40. Oxford, UK: Blackwell.

Lam, N.S.N. Forthcoming. Methodologies for mapping land cover/land use and its change. In *Advances in Land Remote Sensing: System, Modeling, Inversion and Application*, ed. S. Liang. New York: Springer.

Lam, N.S.N, H.L. Qiu, D.A. Quattrochi, and C.W. Emerson. 2002. An evaluation of fractal methods for characterizing image complexity. *Cartography and Geographic Information Science* 29(1):25–25.

Lam, N.S.N., D.A. Quattrochi, H.L. Qiu, and W. Zhao. 1998. Environmental assessment and monitoring with image characterization and modeling system using multiscale remote sensing data. *Applied Geographic Studies* 2(2):77–93.

Lam, N.S.N., C. Catts, D.A. Quattrochi, D. Brown, and R. McMaster. 2004. Scale. In *A Research Agenda for Geographic Information Science*, ed. R. McMaster, L. Usery, 93–128. Boca Raton: CRC Press.

Mandelbrot, B. 1982. *The Fractal Geometry of Nature*. New York: Freeman.

McGarigal, K. 2002. Landscape pattern metrics. In *Encyclopedia of Environmentrics*, Vol. 2, ed. A. H. El-Shaarawi, W. W. Piegorsch, 1135–42. Sussex, UK: Wiley.

McGarigal, K. and B.J. Marks. 1995. FRAGSTATS: Spatial pattern analysis program for quantifying landscape structure. *USDA Forest Service General Technical Report* PNW-351, Portland, OR.

McMaster, R. and L. Usery, eds. 2004. *A Research Agenda for Geographic Information Science*. Boca Raton: CRC Press.

Myint, S., N.S.N. Lam, and J. Tyler. 2004. Wavelet for urban spatial feature discrimination: comparisons with fractals, spatial autocorrelation, and spatial co-occurrence approaches. *Photogrammetric Engineering and Remote Sensing* 70(8):803–12.

O'Neill, R.V., J.R. Krummel, R.H. Gardner, G. Sugihara, B. Jackson, D.L. DeAngelis, B.T. Milne, M.G. Turner, B. Zygmunt, S.W. Christensen, V.H. Dale, and R.L. Graham. 1988. Indices of landscape pattern. *Landscape Ecology* 1:153–62.

O'Neill, R.V., A.R. Johnson, and A.W. King. 1989. A hierarchical framework for the analysis of scale. *Landscape Ecology* 7(1):55–61.

Openshaw, S. 1989. Automating the search for cancer clusters. *Professional Statistician* 8:7–8.

Quattrochi, D.A, N.S.N. Lam, H.L. Qiu, and W. Zhao. 1997. Image characterization and modeling system (ICAMS): A geographic information system for the characterization and modeling of multiscale remote sensing data. In *Scaling in Remote Sensing and GIS*, ed. D. Quattrochi and M. Goodchild, 295–307. Boca Raton: CRC/Lewis Publishers.

Thomas, J.J. and K.A. Cook. 2005. *Illuminating the Path: The Research and Development Agenda for Visual Analytics*. Los Alamitos, CA: IEEE Computer Society.

Woodcock, C.E. and A.H. Strahler. 1987. The factor of scale in remote sensing. *Remote Sensing of Environment* 21:311–32.

Part 3

Visualization and Simulation

8 Spatiotemporal Visualization of Built Environments

Narushige Shiode and Li Yin

CONTENTS

INTRODUCTION

The recent growth in our ability to construct models of cities using a 3D visual representation has been phenomenal (Batty 2000; Batty et al. 2001; Jepson 2006). Thanks to the rapid advancement of information technologies, remote-sensing technologies, online resources, and the increasingly available large-scale spatial data, we see amazingly realistic visual representation of urban environments within which we can immerse ourselves to fly through and interact with the virtual representation of other users as well as the virtual environment itself as if we are in a real city (Jepson and Friedman 1998; Snyder and Jepson 1999). The use of 3D city models as a platform to carry out applications and simulations is also becoming more and more common. In fact, a host of applications and projects have been developed in the planning, designing, decision-making, property marketing, gaming, and other related industries alike (Leavitt 1999; Smith 1999; Padmore 2000; Batty et al. 2001). The range of standards and data formats adopted by these 3D city models is also quite diverse, making some of the models more appropriate for public presentation while other models provide opportunities for in-depth analysis and evaluation of the urban landscape (Shiode 2001; Takase et al. 2003).

One key element is missing from these 3D models, however: the dynamics of the urban growth. As of today, practically all the existing digital models of built environments are static in the sense that they represent the condition of a city only at one specific time point. A city model that can reflect the temporal transition of the urban landscape would not only help us understand the growth dynamics of the urban environment but would also allow us to pursue a whole new range of applications associated with the dynamics of the model. A dynamic 3D city model can also offer an ideal vehicle for promoting spatiotemporal simulations in built environments, both for specific, analytical purposes and for a more general, public representation.

This chapter describes the prospects and issues regarding the designing of a spatiotemporal model of built environments. In particular, it reports on a project aimed at constructing a dynamic model of an urban environment using a 3D GIS platform. Using a case study, we will identify elements that are critical to the 3D GIS modeling of an urban environment, which, in return, would help us refine the methodology prior to wider implementation of further applications such as rapid visualization of built environment for homeland security and monitoring purposes. The project provides critical insights into developing a full-scale, dynamic 3D city model that shows the complete history of a built environment, with the potential to project a possible future landscape through simulation scenarios.

The rest of the chapter consists of the following sections. Section 2 reviews the current range of methods for constructing a 3D city model. It identifies the strengths as well as the limitations of each method and their prospective applications including the planning and management of the urban environment. Particular attention will be paid to (1) the dichotomy between approaches, where one path leads us to the perfection of their visual appearance while the other hones the analytical capability; and (2) the prevailing trend toward online data sharing and Web-based mapping services. This will lead us to the discussion that the current range of city models lack the critical element of temporal dimension.

Following on from the review, Section 3 describes the key concepts that are associated with the spatiotemporal representation of 3D city models. We propose the conceptual framework of two different approaches for constructing such a model. Section 4 describes a case study that uses the methods proposed in Section 3. It describes how prototype models featuring the Buffalo downtown were assembled using different technologies. The prototype, 3D city model is dynamically constructed on a real-time basis, by corresponding to the user's temporal query. Section 5 concludes the chapter with a discussion of the case study findings.

TRENDS OF 3D CITY MODELS

A TREND IN THE 3D CITY VISUALIZATION AND MODELING APPROACHES

Constructing 3D city models presents an iconic and empirical approach to data mining and landscape visualization with a focus on context-specific applications within an urban environment (Day 1994; Delaney 2000). The initial methods of designing such models were based on computer-aided architectural design from an empirical perspective, where detailed measurement of the geometry was regarded as essential. Most of them failed to provide spatial analytical functions, and we had to wait for the emergence of GIS technology (Batty et al. 1999; Environmental Simulation Center 2003). GIS has been intimately associated with our ability to visualize spatial data in maps and related statistical forms, especially for those with irregular and diverse contents such as the urban landscapes (Moltier et al. 2000; Batty and Hudson-Smith 2005).

Increase in the supply of remotely sensed data concerning the 3D environment also helped to make 3D visualization of cities more feasible and popular (Morgan 2000; Teicholz 2000). These activities mainly emerged from developments in geomatic engineering and GIS that met the demands for application of models for querying spatial data structures, and visualizing the results of such queries in the 3D form (Fuchs 1996). The next section shows the range of such data and techniques.

A TYPOLOGY OF DATA AND MODELING METHODS

A number of factors affect the type of 3D urban models we construct, and these include the clients, application, the expected form of output, budget, time period, and the amount of area to be covered. We also find that at least three elements play a critical role in outlining the construction stage of the city models: the degree of reality, types of data input, and the degree of functionality or the ability to conduct various analyses (Batty et al. 2001). The range of variety demonstrated within each factor shows the diversity in the current modeling arena.

The Degree of Reality — The Amount of Geometric Content

The degree of reality is represented by the amount of details captured within the model. Naturally, the more detailed the model becomes, the more cost it incurs. The amount of geometrical details does not necessarily reflect how much reality the model can actually offer, however; in fact, rapid and inexpensive modeling techniques such as texture mapping and panoramic-image capturing have proved to be successful with the general audience (Leavitt 1999). Shiode (2001) proposed a typology of digital representation of cities based on the difference of the geometrical details (Figure 8.1).

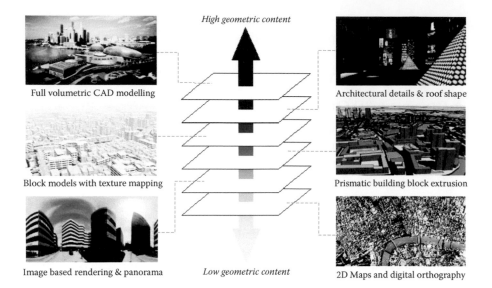

FIGURE 8.1 A typology of 3D city models using different modeling methods (reproduced from Shiode, 2001). **(See color insert after p. 110.)**

1. 2D digital maps and aerial photographs: Conventional 2D GIS maps support a range of applications but are incapable of giving the full 3D representation. However, recent developments of online mapping services, such as Microsoft Live Local, have demonstrated that a series of oblique aerial photographs can be used for enhancing the spatial awareness of its users.

2. Panoramic images: Panoramic image-based rendering is an inexpensive solution to pseudo-3D visualization by offering a 360-degree view of an environment at a fixed viewpoint. The number of viewpoints recorded limits the area it covers, however, and it certainly would not accommodate any analytical capability.

3. Block extrusion: Prismatic building block models are constructed through the combination of 2D building footprints with airborne survey data and other height information (e.g., many of the 3D models found in the Google Earth environment fall into this category). GIS technology allows us to overlay the 2D maps over the airborne data, and determine the spatial characteristics of the image within each building footprint. They lack the architectural detail and are typically monotonous in appearance but are sufficient for certain types of applications, such as calculating the view sheds or determining the shortest route.

4. Block models with textures: This type of model is similar to the prismatic building block model but comes with image-based façades. The building textures are most commonly generated from either oblique aerial or terrestrial images, which, in most cases, seem to successfully compensate for the simple outline of building geometry and roof morphology. There is a natural tradeoff between the quality (resolution) of the image and the data size of the model.

5. Models with some roof morphology: Modern digital photo-grammetric systems enable an efficient recovery of 3D surface details. Automated search techniques are used for identifying the corresponding locations (points, edges, and regions) in multiple, overlapping images to generate a number of possible geometries which can be tested against templates; but still require significant manual intervention for architecturally rich contents. Selected buildings and cities in the Microsoft Virtual Earth environment now have textures mapped to their roof and wall surface, resulting in a fairly realistic representation of the environment.

6. Full volumetric CAD models: As-built CAD models of individual buildings are frequently constructed by combining the data from detailed building survey and terrestrial photogrammetry. The complexity of such models range from digital ortho-photographs (in which images are rectified to remove perspective effects) to the full architectural details that require a significant amount of manual input, and the cost would be prohibitively expensive for a complete coverage of a city.

The level of complexity as well as the time and the cost required for the construction of each type of models applies proportionally to that of a spatiotemporal city model described in later sections.

Types of Data Input — Capturing Heights and Façade Information

The way its data are obtained also affects the final output of the model. For instance, a collection of panoramic images is unsuitable for analytical purposes, whereas airborne survey data provide geometrically accurate but less photorealistic representation. From the spatiotemporal perspective, some data formats would allow partial modification to reflect any updates or changes in the real world, while others require a complete replacement of the entire dataset. Here are some of the data-acquiring methods that are commonly used in the urban modeling context:

1. Terrestrial images: Still images of building façades and video recordings of streetscapes are widely used to provide surface information. It is usually difficult to obtain suitable viewpoints for image acquisition in a downtown district, as the helicopter flight paths and rooftop access would be restricted. In such cases, the building textures are generated from ground-level photographs that often fail to provide optimal façade coverage.

2. Panoramic photographs: Panoramic images provide a highly realistic visualization to all angles from static viewpoints within the survey area. If captured with sufficient density, they can provide a fine representation of an urban area complete with people, vehicles, and street furniture, which are often omitted from the 3D models.

3. Aerial photographs: Remote sensing data have become increasingly accessible and affordable. They provide a rapid and efficient means to cover a wide city area. It should be noted, however, that in order to provide a detailed building façades, a set of oblique aerial images must be acquired

(cf. Microsoft Live Local), rather than the conventional, near-vertical aerial images.

4. Range survey data: Range images can be treated as surfaces over which high-resolution intensity images can be draped, thus enabling the creation of alternative views of the object. The LIDAR (LIght Detection and Ranging) imaging techniques, in particular, are based on camera systems that use a pulsed laser device to record the distance from the camera to each point in the image. Common applications use either ground-based or airborne sensors, the former being suitable for architectural surveys and the latter for small-scale surveys including city models. Airborne LIDAR is invariably used in conjunction with the GPS to deliver high-resolution digital elevation models (Morgan 2000).

As of today, the focus of these data-collection activities is placed on capturing the status quo. They need to be reexamined, if they are to be utilized for the construction of a spatiotemporal city model — especially that of an automatic updating of such model.

Transition I: From CAD to GIS

From the viewpoint of its applications and market demands, the most crucial feature of a 3D city model is arguably that of its functionality (Batty and Smith 2001). Photorealistic CAD-type models are often less functional, whereas the GIS-based models are generally supported by substantial attribute data and offer spatial query and analytical features (Wooley 2000). Here are some of the model types with a different degree of functionality. While the amount of analytical features does not necessarily determine the usefulness of a model in its proprietary context, the potential for extensive and alternative uses will be directly reflected by its functionality, where GIS-based models will prove to be more powerful than their CAD-based counterparts (Liggett and Jepson 1995; Holtier, Steadman, and Smith 2000).

1. Aesthetic models: Models that are intended for aesthetic appreciation and demonstrative purposes generally have little analytical functionality irrespective of their degree of reality. They are designed, generally through the use of a CAD tool, to illustrate the plans and developments to the authorities, various neighborhood groups, or the customers in general through 3D visual representation.

2. Proprietary models with partial analytical features: Typically, a model is equipped with at least one or more analytical features to serve its purpose. These include view-shed analysis, movable buildings, and basic queries that are performed through its built-in analytical tools. These models are usually designed to be self-complete and have little room for extensions.

3. General analytical models: Models extruded from a 2D-GIS dataset often benefit from the use of the full GIS capabilities by inheriting the attribute data attached to the spatial objects within the initial model. The ranges of functions that can be performed in these models include multiple spatial

queries, view-shed and shadow analysis, and various scenario-based simulations; and these models are further extendable in terms of the analytical functions they can offer.

There was a clear separation of models between the aesthetic-CAD-type models and the analytical-GIS types until the early '90s, but this is rapidly changing as the GIS environment is increasingly capable of handling finer and topologically complex models.

Transition II: From Local to Global

The transition from the CAD-based to the GIS-based approach in the arena of the 3D city modeling community became evident in the mid-1990s. In addition to this, there has been a rapid growth in the range of online 3D mapping services that are available on the Web (Doyle et al. 1998).

While the basic online mapping technology itself stems from query-based online services that provide route search function on the street network of a 2D map, their interface is starting to incorporate 2.5D or 3D-like representation. The range of services include Google Earth, Yahoo Maps, and Microsoft Virtual Earth, as well as other customized services available from independent developers through podcasts and blogs.

The foremost advantage of using an online model is that we can share the same environment with other users at the same time (Smith 2000). Allowing multiple users to interact with the same model would widen the range of applications these 3D models can offer. Currently, these environments consist of a combination of different types of city models that were developed separately; thus the degree of accuracy, reliability, and the age of the model varies from one city to another, even within a single environment of Google Earth. This clearly undermines the reliability and quality of such environment. Also, as it is aimed at online distribution, the size of the data is restricted. As a result, the resolution of the surface texture used in these models is relatively low; making these models less appropriate for use in detailed assessment of urban landscape, or for a designing project.

SPATIOTEMPORAL VISUALIZATION OF A CITY

Significance of Spatiotemporal Visualization

This study will offer an empirical understanding of the prospects and issues that arise when combining different types of 3D city models — that is, constructing a model that allows us to visualize and appreciate the growth and change in an urban environment over a period of time.

The transition from the CAD-based models to the GIS approach and the further development of the online city models have greatly improved the utility of 3D city models (Environmental Simulation Center 2003). The diversity of methods applied for collecting the relevant data as well as those for modeling such environment indicate the strong demand for 3D city models in a variety of fields and for applications (Jepson 2006). However, the existing models still differ from the real cities in that they represent townscapes from a single time point. This is not to say that such city models do not accommodate the movement of pedestrians or the interaction among the users, but the urban environment within which they move around remains largely static. There are also some virtual environments such as AlphaWorld or Second Life, which allow their users to "build" and "modify" the representation of their urban environment. They are essentially fantasy worlds and are not a reproduction of an existing city from the real world, however. With the exception of very few examples, such as the Virtual Kyoto project (Yano et al. 2003), no previous attempt has been made on constructing a city model with the spatiotemporal function.

Visualizing the dynamic transition of an urban landscape will allow us to interpret the growth and change of a city from the geographical, planning, economic, and sociological, as well as historical and archaeological viewpoints in an intuitive fashion. It would be particularly useful for understanding the changes in its land-use pattern, residential growth, and many other urban activities. Also, if we can construct such a model in a GIS environment, it would allow us to conduct analysis based on spatioltemporal query. Furthermore, it would open up the possibility for carrying out simulations within the GIS environment to project the future growth of such a city.

In this study, we use two different approaches for constructing a dynamic 3D city model: (a) creating a series of static models from different time points and combining them, and (b) adding a time index to each individual 3D object that is included in the 3D city model.

CONCEPTUAL FRAMEWORK FOR PATCHING MULTIPLE CITY MODELS

The first approach is an extension of the existing method, where a series of city models are constructed, each of which represents a different point in time. By switching between the city models and overlaying one model over another, we can visualize the changes in the urban landscape. It is relatively easy to build, as each city model will be constructed in the same fashion as are the current models. It is also an effective method, even if we have very limited information regarding the year of construction for the buildings.

By reproducing the urban landscape from a selected set of years for which we have the data, we can avoid introducing ambiguity into the model. This approach has the following shortfalls, however: (1) constructing a series of models may require more time and effort; and (2) it also lacks the capacity to handle spatial query across the timeline.

Building a city model for each year would be an enormous task, as it would, in the worst-case scenario, multiply the required time and resource for the construction by the number of time points that we try to represent with the model. In practice, it would require less time than simply multiplying what it takes to build a model of a single year, as many of the buildings would stay for some time and their models can be carried over to other time points. It still takes a lot longer than building a standard city model, however, especially if we try to construct models for many time points. On the other hand, if the model consists of only a handful of time points, the temporal transition of the urban landscape will become more and more discontinuous and abrupt, as each model will have to reflect any changes in the built environment since the status represented in the previous time point.

Furthermore, this type of spatiotemporal model is a compilation of multiple city models, each of which represents the same city but from a different time period. This restricts our ability to conduct spatiotemporal query across different time points. As the model is a collection of separate elements, we cannot conduct analysis that extends beyond each separate city model. This is a major issue, as one of the objectives to create a spatiotemporal model is to enhance our analytical capability for temporal query. Nevertheless, the construction process for this type of model would be relatively straightforward. For instance, we can digitize a series of aerial photographs, which would be an accurate depiction of the urban landscape from each time point.

CONCEPTUAL FRAMEWORK FOR CREATING A SPATIOTEMPORAL DATABASE

The second approach to the construction of a spatiotemporal city model is to make each building object temporally dynamic — that is, to add a temporal index (a time stamp that shows the year of construction and the year of demolition) to each of the buildings that are registered as 3D objects. The basic idea is to create a city model within the GIS environment, on a real-time basis, whenever a user asks for a specific time point. As the user changes the year of selection, the database is searched dynamically and the result will be reflected on the model nearly instantaneously. This approach allows us to dynamically create, through a temporal query, the 3D city model of the year in question. It also allows us to perform other temporal and spatiotemporal queries as we would in a standard 3D GIS city model. For instance, we can perform a query that seeks buildings that were constructed prior to 1950 and are still in existence. We can also select and compare buildings that were built on the same land parcel but at a different time point.

The problem is that many of the buildings would, in fact, be missing information on their years of construction and demolition, which gives the model some element of ambiguity. We thus need to establish a method for visualizing uncertainty, assign a probable temporal index value, and leave it as is. In the case of our case study, we simply estimated the probable years of construction and demolition for buildings that were missing from the land-parcel data, from the aerial photographs, and from the years when similar buildings were constructed in the same neighborhood.

SPATIOTEMPORAL 3D CITY MODEL OF BUFFALO

Using each of the two approaches described above, we carried out a pilot study for constructing a spatiotemporal 3D city model and thereby to examine the validity of such methods. Our study area was a nine-block-wide area around Niagara Square in the downtown of Buffalo, New York.

PATCHING MULTIPLE CITY MODELS OF BUFFALO

The first phase of our study focused on the construction of multiple 3D city models of Buffalo, each of which represented a single time point. This phase consisted of the following four steps:

FIGURE 8.2   An illustrative example of urban transformation in the downtown area of Buffalo: from the small, finer buildings to the larger superstructures (the three snapshots show a portion of aerial photographs of the same location recorded in 1927, 1966, 1978, respectively).

1. Collecting data for five different time points.
2. Digitizing building footprints for each of the five different time points in AutoCAD.
3. Georeferencing the building in SketchUp and adding texture images on the façade.
4. Importing the 3D models to ArcGIS and adding attributes to each feature.

Five aerial photographs from 1927, 1950, 1965, 1978, and 2000 were collected for the purpose of tracing the building footprints. Sanborn fire insurance maps from 1925, 1951, and 1975 were used to confirm the building heights in each model. After some preliminary study, it became clear that buildings in downtown Buffalo were, in general, smaller in units and greater in number in the early periods; they were converted to a small number of large buildings over time. Figure 8.2 illustrates an example of such transition from the smaller, finer buildings of the earlier period to larger, superblock constructions in recent years.

A rough breakdown of the development pattern of Buffalo is as follows: (1) 1880s–1940s: high-density development illustrated by fine-grain building footprints; (2) 1950s–1960s: replacement of the fine-grain buildings with parking lots and mesoscale structure; and (3) 1970s–present: construction of megastructures and superblocks extending over some of the side streets. We thus decided to trace the building footprints for the oldest model of 1927 first, and then use it as our base map for other models so that we would start with a large number of small buildings and then consolidate or replace them with a small number of large building blocks.

Using AutoCAD, the footprints for models of five historical time points were thus digitized from the aerial photograph of the relevant period. Once the building footprints were completed for the 1927 model, we created the city model of the subsequent periods by: (1) overlaying the model from the earlier year on top of the aerial photograph, (2) erasing lines or polygons from the earlier model to reflect any demolition of the buildings, (3) adding features for the new buildings, and (4) extruding the building to the appropriate height confirmed through Sanborn maps. Shadow analysis was also applied on the aerial photographs to verify the building heights extracted from the Sanborn maps. For the 2000 model, building footprints were constructed with an ortho-photograph from 2000 and the current parcel data of the building footprints created in 2002 (Urban Design Project 2003).

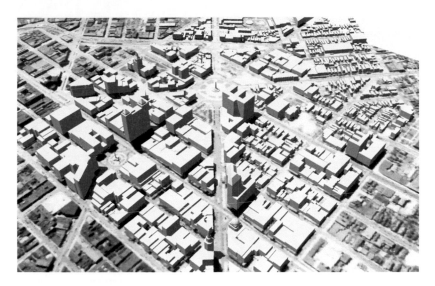

FIGURE 8.3 A snapshot from the 3D city model of Buffalo constructed for year 1927.

Once the footprint files were created for all five time points, we georeferenced them individually and created shapefiles, one shapefile per time point, in GIS. Two attributes were added to each shapefile to indicate the building height and the number of stories. This was completed with reference to the Sanborn maps and also with height confirmation through the shadows shown in aerial photographs.

Figures 8.3–8.5 show snapshots taken from three different 3D city models of Buffalo, each representing a different point in time: 1927, 1966, and 2000, respectively.

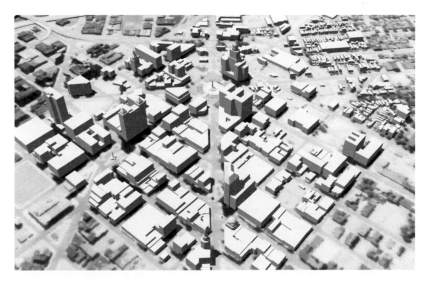

FIGURE 8.4 A snapshot from the 3D city model of Buffalo constructed for year 1966.

FIGURE 8.5   A snapshot from the 3D city model of Buffalo constructed for year 2000.

BUILDING A 3D TEMPORAL GIS MODEL OF BUFFALO

In the second phase of our study, we constructed a prototype of the 3D temporal GIS model that would allow its user to query the spatiotemporal database for the city model.

First, the five city models individually prepared were imported to ArcGIS as a multipatch feature class. A join was performed to get attributes of the building footprint data into the multipatch feature class. To create a 3D temporal GIS model, we combined all the buildings from the five time points into one single model. Besides the height attribute and the number of stories, two more attributes were added to the record of each building: namely, the year of construction, and the year of demolition. In building the 3D temporal GIS model, we considered each building as an individual object with those four attributes attached to it. For buildings that existed through more than one time point, we deleted the redundant one(s), and kept only one 3D object and its corresponding record in the attribute table. In case of the buildings that still existed as of 2000, a large value of 9999 was entered as its year of demolition. If several buildings were associated with the same site, we used the year of construction and the year of demolition to indicate when the building existed on that site.

As the attribute table takes the standard form of relational database used in GIS, we could add other attributes, such as the name of the building, if any, land parcel ID, and so on. This allowed us to link the data of our model with other GIS data of the area, namely the record of buildings in each land parcel. Even after we joined the land parcel data, the construction year and the demolition year were still unclear for some of the buildings. In such cases, the Sanborn maps and the aerial photographs were used to estimate those years — that is, if a building was present in the 1927 map but not in the 1955 map, we assumed that it was demolished in 1954. Clearly, this method does not give us an accurate account of townscape for each year. It does provide, however, a rough estimate of the construction activities, sufficient for our purpose of pilot testing the validity of our method to construct such a GIS model.

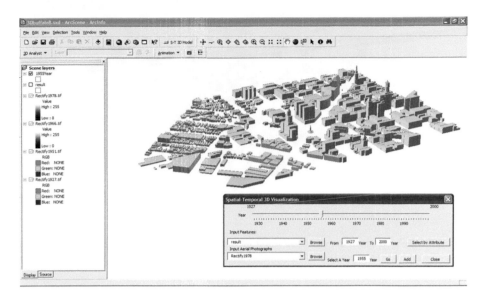

FIGURE 8.6 A prototype interface for the dynamic 3D city model showing the downtown area of Buffalo from 1955. The space-time scroll bar in the lower-right corner controls the representation of temporary dynamic 3D GIS model and allows its user to select a specific year or attribute to be visualized in the main window.

INTERFACE OF THE SPATIOLTEMPORAL MODEL

One of the advantages of using GIS as a platform is the availability of the various spatial query and display functions for selected attributes and objects. Yet, most of the current range of GIS software does not provide a tool for querying 3D objects. In this study, we designed a custom interface to conduct the following functions: (1) Select a specific year for querying the GIS model; (2) make a selection by the attributes associated with each 3D object; and (3) display only the selected 3D objects, either by the contemporaneous state, or other selected attributes.

We created an additional interface as shown in Figure 8.6, which allows the users to perform these functions. A temporal scroll bar was also added so that the users could move the time bar to a certain time point and view all of the buildings that existed at that time point, as well as explore the attributes of those buildings. As the user slides the time bar to select a different time point, the query mechanism would search for the 3D objects (buildings) that existed at that particular point in time, and dynamically generate a layer of 3D objects each time. The result was a truly interactive spatiotemporal city model that allowed us to visually appreciate the dynamic growth of an urban environment. Figure 8.7 shows an illustrative example of how the spatiotemporal city model of Buffalo can be used for a temporal query. The darkness of the shade used for the buildings represents the years they were constructed— that is, the darkest ones dating back to the 1840s, while the lightly shaded ones were built after 1966.

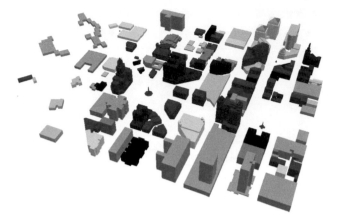

FIGURE 8.7   A close-up view of the 3D model of downtown Buffalo, where its buildings are shaded by the temporal query on their year of construction. Buildings in darker shade are older construction, and those in lighter shade are more recent.

CONCLUSION

SUMMARY

This chapter discussed the prospect of building a spatiotemporally dynamic 3D city model in a GIS environment. Using a prototype city model, we proposed and studied a framework for constructing a fully dynamic 3D representation of an urban environment that can be used in a variety of analytical, planning, and simulation contexts.

In order to visualize the urban landscape from different time points, we introduced two different approaches. The first approach was to construct a series of city models from different time points and combine them as multiple layers within the GIS environment. The second approach was to embed the temporal information, or the time stamps, into each building object, and thereby, to allow the users to construct the model dynamically through temporal query. Although the first approach was easier to achieve, it took considerably more time and more effort to construct each individual city model from different time points. In contrast, the second approach demanded less time for the model construction, but required us to design an entirely new system and interface for the real-time construction of the spatiotemporal city model. It was one of the first attempts at designing such a dynamic system for visualizing a 3D city model in the GIS environment.

ISSUES IDENTIFIED IN THE CASE STUDY

In order to verify the validity of the two approaches, a pilot study was carried out on a small segment of the downtown area of Buffalo, New York. While the case study successfully demonstrated the feasibility of the two approaches, it also raised several issues regarding the difficulties and limitations of designing such a model.

For instance, during the case study, we found that many buildings were missing the basic information about their year of construction and year of demolition, as well as their height and the materials used. The missing data were filled by estimated values through a close examination of the historic aerial photographs and the fire-insurance maps, but the margin of error was occasionally as big as a decade. This could be a much more significant issue if the city has existed for many centuries before the time of aerial photographs. In our case study, we simply adopted those values, but such ambiguity should be treated with caution and may require the adoption of a different representation in the future — for example, a semitransparent appearance for the buildings that are represented during its transitional stage — which unfortunately was not feasible in the current GIS settings. Visualization, representation, and interpretation of a spatially-temporally uncertain data are yet to be established and may indeed require a new framework.

Also, as standard GIS environments are not designed for the temporal representation of an urban landscape, we had to develop our own proprietary interface. During the testing phase of the case study, the layout of this interface saw several modifications to incorporate feedback from our test team. The whole system is yet to receive a formal evaluation, however.

Finally, the prototype interface we developed is still bound by the rules of the current GIS framework, and it cannot be utilized online, or applied to spatial-temporal data of other cities that do not conform to such framework.

FUTURE PROSPECTS

The technology proposed in this chapter is intended primarily for developing a dynamic landscape of urban environment. It is fundamentally applicable to constructing other terrains and environments, however. For instance, we can, with some adjustments, use it for representing the change in the physical or natural landscape that would usually occur on a much longer time scale.

This chapter proposed methods to visually represent an iconic model of a city from different time points. It provides the first step toward designing such a spatial-temporal model and requires further research toward its improvement. The next steps are to address some of the issues stated above, apply the spatial-temporal interface to a larger dataset, and add more analytical functions for simulation purposes (e.g., extracting the pattern of urban growth through the change in the location of centroids of the buildings, and projecting the future growth).

REFERENCES

Batty, M. 2000. The new geography of the third dimension. *Environment and Planning B: Planning and Design* 27:483–84.

Batty, M., D. Chapman, S. Evans, M. Haklay, S. Küppers, N. Shiode, A. Smith, and P. M. Torrens. 2001. Visualizing the city: Communicating urban design to planners and decision-makers. In *Planning Support Systems*, ed. R. Brail and R. Klosterman. New Brunswick, NJ: ESRI Press and Center Urban Policy Research, Rutgers University Press.

Batty, M., M. Dodge, B. Jiang, and A. Smith. 1999. Geographical information systems and urban design. In *Geographical Information and Planning*, ed. J. Stillwell, S. Geertman, and S. Openshaw, 43–56. Heidelberg, Germany: Springer Verlag.

Batty, M. and A. Smith. 2001. Virtuality and cities: definitions, geographies, designs. In *Virtual Reality in Geography*, ed. P. F. Fisher and D. B. Unwin. London: Taylor and Francis.

Batty, M. and A. Hudson-Smith. 2005. *Imagining the Recursive City: Explorations in Urban Simulacra*, CASA Working Paper 98, University College London.

Day, A. 1994. From map to model. *Design Studies* 15:366–84.

Delaney, B. 2000. Visualization in urban planning: They didn't build LA in a day. *IEEE Computer Graphics and Applications*, May/June 2000: 10–16.

Doyle, S., M. Dodge, and A. Smith. 1998. The potential of Web based mapping and virtual reality technologies for modeling urban environments. *Computers, Environments and Urban Systems* 22:137–55.

Environmental Simulation Center 2003. *Combining 3D modeling with GIS*, Environmental Simulation Center, New York (http://www.wenet.net/~shprice/Kwart1.htm).

Fuchs, C. 1996. OEEPE Study on 3D-City Models. *Proceedings of the Workshop on 3D-City Models*, OEEPE (Organisation Europeenne d'Etudes Photgrammetriques Experimentales), Institute for Photogrammetry, University of Bonn, Bonn, Germany, 37 pp. (with appendices).

Holtier, S., J. P. Steadman, and M. G. Smith. 2000. Three-dimensional representation of urban built form in a GIS. *Environment and Planning B: Planning and Design* 27:51–72.

Jepson, W. 2006. A real-time visualization system for large scale urban environments. School of Architecture, UCLA, Los Angeles, CA (http://www.aud.ucla.edu/~bill/UST.html).

Jepson, W. and S. Friedman. 1998. It's a bird, it's a plane, it's a supersystem: How all of Los Angeles is being captured on computer. *APA Planning* 64(7):4–7.

Leavitt, N. 1999. Online 3D: Still waiting after all these years. *Computer*, July 1999: 4–7.

Liggett, R. and W. Jepson. 1995. An integrated environment for urban simulation. *Environment and Planning B* 22:291–302.

Morgan, B. A. 2000. Evaluation of LIDAR Data for 3D-City Modelling. MSc thesis, University of London.

Padmore, K. 2000. The Liverpool Project, The Centre for Virtual Environments, University of Salford, Manchester, UK.

Shiode, N. 2001. 3D urban models: recent developments in the digital modelling of urban environments in three-dimensions. *GeoJournal* 52(3):263–69.

Smith, A. 2000. Shared architecture: rapid-modeling techniques for distribution via on-line multi user environments. *Arcadia* 19(1), in press.

Smith, S. 1999. Urban Simulation: Cities of the Future, A/E/C/Systems (Architecture, Engineering, and Construction Automation).

Snyder, L. and W. Jepson. 1999. Real-time visual simulation as an interactive design tool. Paper presented at the ACADIA 99 Conference, Snowbird, Utah.

Takase, Y., N. Sho, A. Sone, and K. Shimiya. 2003. Generation of digital city model. *Journal of the Visualization Society of Japan* 23(8):21–27.

Teicholz, N. 2000. Shaping cities: Pixels to bricks. *New York Times,* Technology Circuits, Thursday, December 16, 1999.

Urban Design Project. (2003). *The Queen City hub: A regional action plan for downtown Buffalo*, City of Buffalo, New York. Also available from http://www.urbandesignproject.org/.

Wooley, K. 2000. Photorealistic imaging of GIS data. *Geoinformatics* 3:12–15.

Yano, K., Nakaya T., Isoda, Y., and Y. Takase. 2003. Virtual Kyoto: Restoring historical urban landscapes using VR technologies. *Paper presented at the 2nd NII International Symposium "Digital Silk-roads,"* Nara, Japan.

9 Visual Analysis of Urban Terrain Dynamics

Thomas Butkiewicz, Remco Chang,
William Ribarsky, and Zachary Wartell

CONTENTS

INTRODUCTION

Modern urban areas are places of continuous change. Over months, buildings may be torn down and new ones started; streets can be altered and new ones constructed; railways or other means of urban transport may change. The models of urban areas must be able to accommodate these changes. This is especially so since models are significantly higher resolution than previously and cover wider areas. Now models can typically have imagery and elevation data at resolutions of one foot to one meter with certain features (on key buildings, for example) that may be at higher resolution. At these resolutions and for certain applications, even small changes can be noteworthy.

In addition, urban terrain models can come from many sources. These include varieties of sensors such as LIDAR, satellite imagery, airborne oblique photography, ground-based depth and appearance fields, SAR, and so on. Automated, semiautomated, or manual techniques are used to reconstruct the urban model. In the latter case, users may use 3D design software to create individual building or streetscape models from combinations of photographs, measurements, and building plans. A comprehensive, dynamic model should be able to handle contributions from any and all of these sources. In many cases, such as urban planning, civil engineering, or military applications, a lower-resolution model of an urban area of interest will be augmented with higher resolution data, which may come from a sensing source other than the original data. These data must be embedded into the context of the existing model, and often time is not available (nor for certain applications, should it be necessary) to reconstruct the whole model based on the new data.

A great deal of work has been done to develop interactive visualization and terrain analysis methods for large-scale, high-resolution terrain. Most of these methods, however, treat the terrain model as a 2D surface, and in many cases just as a height field. But for current urban terrain applications, a significant need exists to treat the terrain as a 3D model with multivalent heights and non-genus-0 topologies. Overhangs, subways, subterranean rooms and passages, bridges, and so on are all of interest in these models. In addition, the variety of data sources means that a comprehensive modeling approach must deal with overlapping patches or volumes from different sensors collected at different times. The modeler must deal with how to use this combination of resources (e.g., does one merge based on an analysis of overlapping patches, choose the most recent or "best" patches, etc.), deal with missing data and error, and keep track of the temporal history of the evolving terrain.

With respect to this last point, certain applications need to treat urban terrains as models with significantly faster dynamics. Military applications for urban combat zones, for example, must consider sudden damage to or complete destruction of buildings, roads, bridges, and so on, as well as cratering of the terrain surface and collapsing of subterranean structures.

In this paper, we present the initial steps in a comprehensive approach to organizing and using large-scale, high-resolution, and dynamic 3D urban terrains. This approach can incorporate terrain data from all sources, including those described above, in the form of meshes, sampled point clouds, depth and appearance images, implicit surface models, volumetric models, and others. Errors, uncertainties, and confidence measures, both for the terrain models and for analyses based on the models, can be propagated in a multiresolution, hierarchical structure. The different types of dynamic effects described above can be handled in an efficient and effective way. A hallmark of this approach is fast update in localized regions where dynamic changes take place without impacting the rest of the representation.

In this paper we also focus on two types of applications: terrain analyses such as line-of-sight, trafficability, and penetrability; and urban planning involving both long-term planning and large-scale urban projects. The former application is of prime interest to the military but is also of interest to engineers and others. The latter application addresses issues such as dynamic zoning and how large-scale projects fit in the overall plan, which are issues increasingly important to city planners.

COMPREHENSIVE VOLUMETRIC APPROACH

Our approach embodies not only the sample points, resulting mesh, and so on that describe the terrain, but also the inherent idiosyncrasies and shortcomings that are characteristic of the various methods used to collect terrain samplings. In addition, the geologic qualities of the terrain itself (such as surface composition, roughness, bogginess, etc.) are also taken into account during the calculation of the final terrain models derived from samplings. For generality, we develop a volumetric representation for the terrain, which embeds the uncertainty/error from both the sampling techniques and the terrain's physical qualities. The final volumetric representation, essentially formed by upper and lower bounds, can then be considered to encompass all of the possible physical surfaces that could have resulted in the original set of samples. It can also be extended to encapsulate different types of samplings from different sources (e.g., different sensors) into a single comprehensive representation. Finally, the representation can be extended to include fully volumetric terrain, as will be discussed further below.

The upper and lower bounds of our volumetric representation, derived from sampling error and geologic variations (for the given terrain type), can be determined for models that are connected or not, regularly or irregularly sampled, or that have multivalued heights or full 3D structure.

Our approach for transforming the sample data into a volumetric representation is based on the voxelization techniques in Zelinka's Permission Grids (Zelinka and Garland 2002). Zelinka uses this approach to provide a precise upper bound for multiresolution mesh simplifications. We have extended the approach to families of terrain models with their own characteristic errors and geologic variations. In Zelinka's original algorithm, a volume is created entirely within a static specified distance (ε) of the original triangular mesh. We have replaced this static ε with a dynamic, nonuniform distance metric that adapts to each location on the terrain. For each sample point in our data, we apply our sampling error metric to create a volume around each sample point, bounded by the positional, elevation, and other errors for our desired confidence level. In order to compute confidence levels, ε is specified in terms of characteristics of the error distribution. (For example, ε is specified as the 1σ or 2σ distance for a Gaussian distribution.) In general, ε is a vector since the error will be asymmetric. (With LIDAR, for example, the error is significantly larger in the lateral direction than in the vertical direction.) Thus, for greater efficiency, our voxels can be noncubic.

For all regions between sample points, we combine both the sampling error and the possible geologic error, discussed below, to determine values for ε at all points on the terrain. This allows us to fill in the volumetric model between the available sample points. By understanding the limitations and errors inherent in a sampling technology, we can achieve a volume that reflects a desired confidence level and thus can be traversed and evaluated very efficiently.

We implement the volume in a 3D data structure similar to an octree. Final minimum voxel size can be chosen at runtime and is limited based on available computational and storage resources. The ratios of the minimum voxel size and different values of ε determine the precision of the volume. By decreasing the minimum size of the voxels we increase the precision of the voxels as approximate fits to the confidence bounds. During the volume-creation phase, we recursively subdivide the volume until it reaches, for each local region, the necessary voxel size. After this process completes, the structure then proceeds to recursively remove redundant detail.

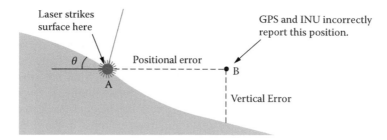

Laser strikes
surface here

GPS and INU incorrectly
report this position.

θ

Positional error

A

B

Vertical Error

FIGURE 9.1   An example of nonuniform sampling error taken from LIDAR. As the slope (θ) of the terrain increases, the error in horizontal position has increasing effect on the accuracy of vertical measurements.

This hierarchical structure permits the creation of multiresolution models for different applications with fast access and minimum storage and memory requirements. The model can also be split apart into smaller models collected from different sources and/or processed in parallel.

SAMPLING ERRORS

Sampling errors, which vary from point to point, depend on the characteristics of the methods used to acquire data and their effects across different terrain regions. For example, LIDAR, which uses a pulsed laser that is scanned from an aircraft, returns a depth component that is quite accurate. However, the horizontal component, generated using a combination of GPS and inertial navigation updates (INU), can be considerably less accurate. In addition, scattering from corners of buildings and other effects can produce significant degradation in the depth reading. Deviation of the model from the actual terrain is also affected by the nature of the terrain itself. For example, as the slope of the surface increases, errors in the horizontal components significantly affect the accuracy of the vertical component. Figure 9.1 shows this effect in detail. Our approach was specifically designed to account for these types of errors.

GEOLOGIC VARIATIONS

In cases where surface models are tessellated, the areas between sample points are linearly interpolated. Such an approximation is more or less accurate depending on the nature of the underlying terrain. If the sampled terrain is prairie and grassland or a smoothed city terrain (Figure 9.2a), we would say that the possible vertical difference (geologic variation) of the actual terrain from the linear interpolation between sample points would be rather small. However, if the sampled terrain is craggy and prone to unpredictable protrusions and pits, such as shown in Figure 9.2b, the geologic variation between points sampled at the same density could potentially differ quite significantly from the interpolation between the sample points.

Terrain cover, such as tall grass or other vegetation, could also make a difference for certain applications. One can obtain the size and nature of these effects from GIS layers giving the terrain type and properties and also from statistical evaluations of

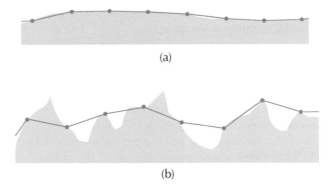

FIGURE 9.2 (a) For a grassland terrain, the linearly interpolated surface between the sampled points varies only slightly from the actual terrain, while for a rocky terrain, (b) the outcroppings between sample points protrude far past the interpolated surface.

terrain variability in the region of the sample points. Our approach takes account of these factors. We show in Figure 9.3 how total error, which in this case is sampling error plus geologic variation (assumed Gaussian here), is determined. One could have additional errors or more complicated representations.

VOLUMETRIC TERRAIN ANALYSES

These volumetric terrain models are quite suitable for a number of terrain analyses, including line-of-sight, penetrability, and trafficability.

For line-of-sight (LoS) applications, the integrated error bounding allows certainty measures to be tied to visibility calculations. This is of particular importance for time-critical military LoS calculations, where terrain data acquisition is likely to have been rushed, perhaps haphazardly, or even partially incomplete. By placing a conservative bound on the possible errors, we can do more than simply report what regions are visible and invisible to units, but also calculate what regions are of questionable visibility and the degree of certainty as to their status.

Because the terrain model is volumetric, the results do not just indicate the visibility (as areas) on the ground (as most other methods do) but instead indicate visibility volumes. This is important when dealing with the hiding or discovery of

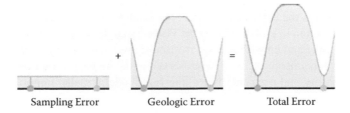

Sampling Error   Geologic Error   Total Error

FIGURE 9.3 The total error is a combination of the error resulting from the sampling processes and the error due to geologic variations between sample points. The shaded areas underneath the curves depict the bounded error volume that encapsulates the surface.

units/objects of substantial size or those in flight. An example of a situation where this is useful is the calculation for an optimal flight path of an unmanned aerial vehicle over a combat zone that will allow it to produce useful reconnaissance of unobserved areas and possible vehicle movement, while avoiding visual detection from known enemy positions.

Another benefit of a volumetric representation (over 2.5D methods) is that instead of being limited to calculating visibility from single eye-points (i.e., "point-to-point" and "point-to-area"), the user can do "volume-to-volume" calculations. This permits the calculation of the visibility of a unit's entire patrol area, a group of units, or a complex with multiple observation points. Figure 9.4 shows this capability and Figure 9.6 shows the underlying multiresolution data structures.

The multiresolution and hierarchical nature of our volumetric terrain models permits our applications to dynamically adjust the balance between accuracy and speed of calculations. When calculation time is not an issue, the system can use the highest resolution terrain data for maximal accuracy and confidence in the results. In a time-critical situation where the user desires a result quickly, however, the applications can lower the resolution of the terrain and calculate orders of magnitude faster. Because each resolution level has an inherent confidence level associated with it, the applications can inform the user just how inaccurate these "rushed results" may be. Conversely, the user can specify a bound on the confidence and the system can adaptively switch to resolutions that provide the desired accuracy while calculation times are kept to a minimum. Because of the ability to maintain local as well as global errors in the terrain model simplifications, the user or automated manager can control where computational effort, and thus accuracy, is concentrated. A simple but effective method for concentrating computational power and accuracy is to define regions-of-interest, shown in Figure 9.4 as yellow boxes of highlighted terrain. Once the user defines a region-of-interest, the system automatically loads in the highest-resolution terrain data it can locate for that particular region and produces a volumetric model of the highest resolution allowable under the current memory and time constraints. These methods are quite effective as shown in Table 9.1 for the scenario in Figure 9.5. Further discussion of this work can be found in Butkiewicz et al. (2007).

When dealing with a large-scale, high-resolution terrain model, one can achieve a significantly higher data quality to storage ratio by identifying features of the terrain that are of importance for a specific application and storing these areas at a higher resolution than the surrounding terrain of less importance. A good example of this concept is the identification and preservation of ridgelines for terrain models that will be used in line-of-sight or other visibility applications. Because the volumes of visibility over and under horizons are almost always determined by the ridgelines of a terrain, it is imperative they be preserved at the highest resolution possible.

Penetrability and permeability of a terrain model are also important analytical concerns. Here *penetrability* refers to the actual physical entrance of the terrain (e.g., through digging or an explosion), whereas *permeability* refers to the ability to see through the terrain (or its foliage) visually or by sensor radiation. Ground cover, primarily vegetation, can be detected by and is somewhat permeable to scanning technologies such as LIDAR, which produces returns both on and in the vegetation or

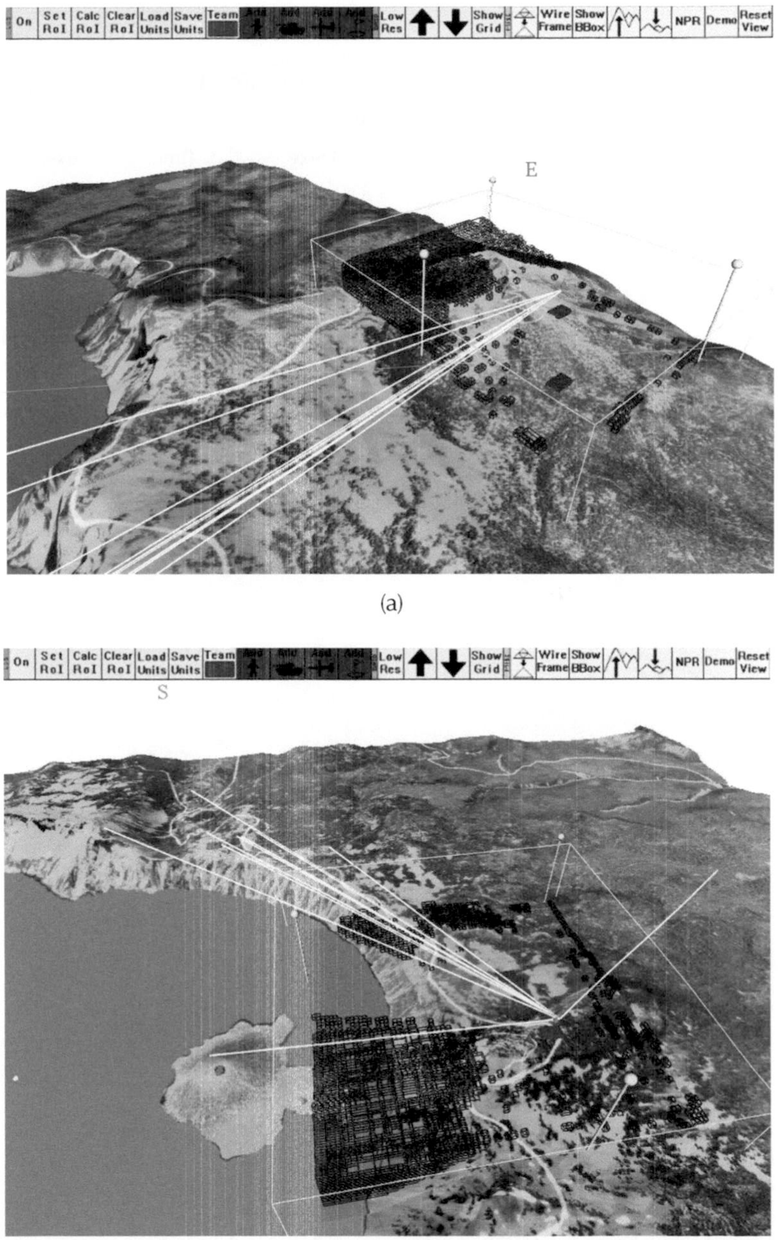

(a)

(b)

FIGURE 9.4 Example results of both point-to-point and point-to-volume visibility calculations in our LoS application. Volumetric results are depicted as black boxes in a yellow region-of-interest. (Note that the voxel sizes here have been enlarged for illustrative purposes.) Point-to-point visibility from the red objects are shown here as connecting lines-of-sight, but can also be represented as icons above visible or invisible units. **(See color insert after p. 110.)**

TABLE 9.1

Statistics for Arbitrary Accuracy Levels Determining Balances between Model Accuracy and Computation Time.[a]

Accuracy Level	Max Error Allowed	Voxel Size	Time	Voxels Traversed
Low	150 m	85.7 m	0.12 s	164,519
Medium	100 m	50.0 m	0.19 s	257,649
High	50 m	16.6 m	0.56 s	716,466
Best	30 m	10.0 m	0.92 s	1,172,485

[a] Calculations were done on a terrain of size 20 km × 14 km generated from a 30 m resolution data source. Establishing visibility information for the two teams (each of 53 units) required 2,809 inter-unit visibility calculations.

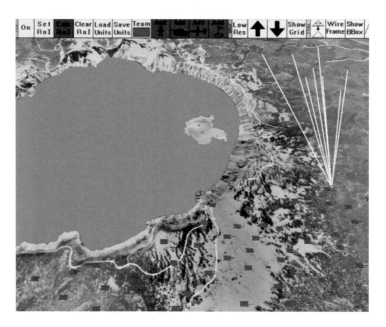

FIGURE 9.5 Line-of-sight scenario consisting of two teams of 53 units each across a 20 km × 14 k terrain. Statistics for this scenario at different levels of accuracy are given in Table 1. **(See color insert after p. 110.)**

canopy and the ground itself. These returns can be classified as canopy, bare ground, buildings, and so on. By treating vegetation cover as a volumetric layer above the ground, and assigning a density to these volumes, we can define the permeability of the vegetation from air to ground.

Various methods of measuring the earth's composition with different radiation bands exist. One such technology is synthetic aperture radar (SAR), currently used to measure terrain structure for purposes that include the study of geological structures such as volcanoes, active faults, landslides, oil fields, and glaciers. SAR that

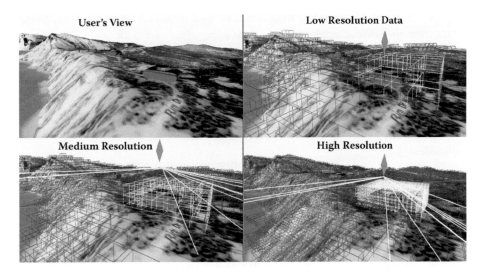

FIGURE 9.6 The user's view presents a simplified terrain mesh, while underneath the calculations are performed on the volumetric models. The application can adaptively switch between multiple resolutions (each with their own confidence measures), maintaining the desired balance between computational speed and accuracy or confidence of the results. **(See color insert after p. 110.**)

maps areas of the earth's surface with resolutions of a few meters can provide information about the nature of the terrain and what is on its surface These data can give insight as to the penetrability of terrain and also for applications such as trafficability or flooding.

This research is of particular interest for military applications. Below-ground structures such as fortified military bunkers, fuel-delivery lines, and utility infrastructure are considered challenging targets, with "bunker-buster" weapons receiving much attention. By studying and understanding the complex nature of both the penetrability and permeability of the earth and what covers it (buildings, vegetation, etc.), one can make tactical and strategic decisions about these traditionally difficult targets. We are now extending the 3D volumetric approach presented here to address both surface and 3D terrain models, traditional buildings, and subterranean structures.

Dynamic Terrain

It is most desirable to extend past a static single-sourced terrain model, utilizing multi-data-sources for the creation and maintenance of a comprehensive terrain representation. The terrain and structural elements of large population centers receive frequent sampling and scanning from LIDAR, satellite photography, and other techniques. Databases of building footprints and models are constantly updated for insurance and tax purposes. It is crucial to develop a terrain system that is not only capable of recognizing and integrating as many diverse, overlapping datasets as possible but that also possesses an understanding of both the age of and errors present in each source.

A system that allows fast integration of new data and removal of out-of-date data from the current amalgamated model is necessary for wide-scale terrain models where single data sources are insufficient and for terrains that are modified or resampled often. Construction sites for new residential developments can replace wilderness and can then be replaced with the final buildings on completion. Military commanders need to be able to easily remove destroyed buildings or add craters to their models so that changes can be immediately propagated to ground units. Thus we need to be able to handle terrain dynamics on scales of days to hours, especially for high-resolution terrain.

Terrain models that contain temporal history are also important for a number of reasons. For example, they can enable city planners and historians to view county-wide areas and visualize changes and development over time. Developments can be evaluated for their environmental or aesthetic impact on the surrounding land. The storage and retrieval process requires spatiotemporal access methods (STAM). The STAM research is found in many disciplines, including databases (with subspecial-ties in temporal [Zaniolo et al. 1997], spatial, and spatiotemporal [Bohlen et al. 1999] databases); GIS and computerized cartography (Voisard et al. 2002); and computer graphics and visualization (Shen et al. 1999). Our approach is to develop an event- or feature-based structure based on the concept of interval trees (Edelsbrunner 1980). Features could be ridgelines, subterranean structures, the urban morphology fea-tures described below, or any structure of importance. Changes in these structures can be followed over time with significant changes (events) noted. The events are then visualized along a timeline, with associated lower-level features at any level of detail required. This event structure is typically much smaller than the original data and may in certain cases be orders of magnitude smaller. The interval tree then permits efficient retrieval of full data when needed.

TESTBED DATASET

Large-scale, specific, and comprehensive terrain data are hard to obtain. There must be data from several sources, including multiple types of capture and modeling. There must also be 3D data, data subsets that can be added or deleted, and regions where different sources cover the same area (sometimes in contradiction with one another). To ensure having data freely available for our research, evaluation, and application development, we have our own multisource, dynamic dataset. We plan to continue adding to this dataset to make a comprehensive source. This dataset will be used for evaluation and to try out ideas on data organization, terrain analyses, and so on, and will be shared freely with collaborators and other interested parties.

Our initial testbed dataset is a collection of data sources covering the entire county of Mecklenburg, NC (where Charlotte is located). (See the excerpt in Figure 9.7.) Ter-rain data are from three sources: USGS topological surveys in DEM format at 30 m resolution, a more recent (2003) DEM at 20-foot resolution, and a 20000×20000 (~4 m resolution) LIDAR return collection. The LIDAR returns, which cover the entire county, are also classified (building, vegetation, bare earth, etc.). While buildings are easily distinguishable in the LIDAR returns, we also possess footprint data for all 370,000+ buildings and structures in the county. By comparing these footprints to

FIGURE 9.7 Excerpt from testbed dataset with collection of building models, underlying DEM, and classified LIDAR point cloud. **(See color insert after p. 110.)**

the LIDAR returns, we can associate heights with them and re-create simple models for each. Architecture students have manually modeled landmark buildings, campus buildings, and other structures. To provide 3D test data for subterranean structures, a notional subway system was placed under downtown Charlotte, as well as bunkers and other underground construction, such as utility infrastructure. We have organized the dataset so that different sources can be added or subtracted and, in particular, so that data acquired from real locations can be separated from notional data. This organization will also permit us to study dynamic effects.

REPRESENTING COMPLEX URBAN MODELS USING URBAN LEGIBILITY

Complex urban models can be composed of hundreds of thousands of buildings laid out on high-resolution terrain elevation maps covered with ortho-rectified imagery. The building models can be generated automatically from a combination of footprint and height data (the latter from LIDAR, for example) with generic or more specific textures. More detailed specific buildings would then be generated with semiautomated methods and embedded in the database. Ultimately these models should also be represented in the volumetric approach described in the last section. However, as we will see, if the high-level applications for these models are different, the multi-resolution representations must be different, too.

   Although automatically generated building models are often very simple in geometry due to the fact that they are 2.5D protrusions of footprints, interactive viewing and manipulation of a large number of these buildings can still easily exceed the available memory on most computers and the capabilities of modern graphics cards. For interactive viewing and manipulation of a large number of building models, a simplification scheme is essential. Unfortunately, traditional decimation techniques (Garland and Heckbert 1997) (Luebke 2001) do not work well for simplification of

Original (textured) District   Simplification with our Method

Simplification with Qslim   Our Simplified Model with Texture Applied

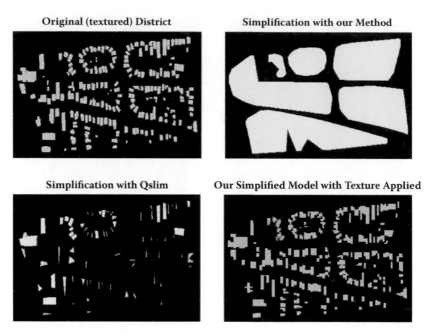

FIGURE 9.8 Comparison of our method (right) to original data (top left) and to a well-known simplification method (Qslim [Garland and Heckbert 1997], bottom left). Both methods have the same number of polygons, but Qslim loses all sense of the original model. After texturing, our greatly simplified model (bottom right) retains the features of the original model.

urban models (see Figure 9.8), as the decimation process often creates models that no longer resemble the originals.

To simplify urban models in a meaningful way, we apply the principles of urban legibility. Urban legibility is a concept that has been used for many years in the area of city planning as well as computer graphics (Dalton 2002) (Ingram and Benford 1995), and is defined as the ease with which parts of the city can be recognized as a coherent pattern and thus contribute to an accurate and usable mental map of the city. In his book *The Image of the City*, Kevin Lynch (Lynch 1960) further categorizes the elements of urban legibility into five groups:

- Paths: streets, walkways, railroads, canals, etc.
- Edges: boundary elements such as shorelines, walls, edges of developments
- Districts: medium to large sections of the city, such as a specific residential district, that have their own existence to the observer
- Nodes: strategic spots of intense activity (e.g., Times Square)
- Landmarks: recognizable structures that are distinct to the observer

To make large-scale urban models interactive and navigable, we aggregate these large models that may contain hundreds of thousands of buildings using the above principles (see Figure 9.9). Hierarchical textures are applied to the aggregated models in a manner similar to imposters (Sillion et al. 1997) (Maciel and Shirley 1995), and

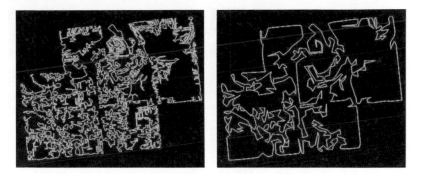

FIGURE 9.9 Outline of an urban model after aggregation. The left image shows the original outline of the aggregated models; the right shows the simplified outline. The simplification process preserves the legibility elements in the city model, thereby creating a simplified model that remains legible and understandable. (**See color insert after p. 110.**)

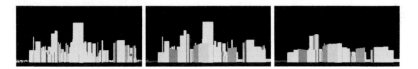

FIGURE 9.10 Maintaining the landmarks and the skyline. The image on the left shows the original model (243,381 polygons), the image in the middle shows a simplified model with landmark preservation (15,826 polygons), and the image on the right shows an unpreserved skyline (13,712 polygons). Note that the quantitative difference between turning on and off landmark preservation is merely 2,114 polygons in model complexity, but the visual difference between the two is very significant.

landmarks are treated with special care so that they are maintained throughout the simplification process in order to retain the skyline (Figure 9.10). The result is a general approach that produces multiresolution models of cities that are legible, understandable, and navigable at all levels of detail (Chang et al. 2006) (see Figure 9.11). Figure 9.12 shows an excerpt from one model at different levels of detail, and Figure 9.13 shows the number of polygons and frame rates using different levels of

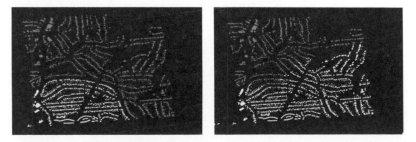

FIGURE 9.11 Clustering buildings in a city. The left image shows clustering results that follow the urban legibility element Paths. The right image shows the result of a more traditional distance based clustering. (**See color insert after p. 110.**)

FIGURE 9.12   Original textured 3D model (left) of Xinxiang, China; simplified model (right) with only 18 percent of the original number of polygons and aggregated textures. View-dependent rendering is applied to the hierarchical multiresolution structure on the right. **(See color insert after p. 110.)**

detail in a flythrough scene. From these two figures, we see that by using urban legibility as the basis of urban model simplifications, we can achieve drastic increase in rendering speed by decreasing large numbers of polygons without sacrificing too much visual realism. Within the context of dynamic terrain, we can now consider the factors below.

DISCOVERING ELEMENTS OF LEGIBILITY

Using our methods, we can create clusterings and groupings of urban models such that the groupings obey the general rules of urban legibility. This doesn't actually identify the exact legibility elements themselves that define a particular urban model,

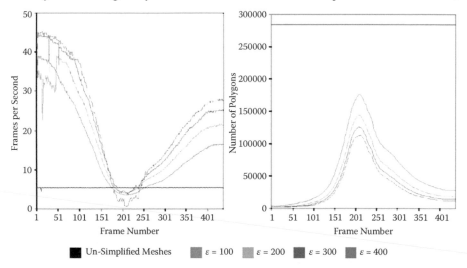

FIGURE 9.13   Frame rate and polygon counts. Using simplified models with different levels of detail in a flythrough scene (a high ε value denotes a large amount of simplification), we see that our simplification method can provide drastic speedup by decreasing the number of polygons rendered to the screen. **(See color insert after p. 110.)**

FIGURE 9.14 Finding elements of urban legibility. To correctly identify the elements of urban legibility and hierarchically categorize them from most to least important, we overlay different levels of detail of the urban model over GIS data that contain information on roads, districts, etc. The images from left to right show the urban model in decreasing LoD. The coarsest level of the model retains the most significant elements in the city, such as the waterway that cuts through the city diagonally, the park in the center of the city, and the main road that runs east-west through the city; whereas the finest level of detail contain much finer elements that are only significant in a localized way.

however. Finding out which of the five urban legibility elements are creating the logical districts and groupings is important for both theoretical and practical purposes. From a theoretical standpoint, finding the elements of legibility can help urban planners quantify the major features of a city. From a practical standpoint, knowing which elements are important in a city can assist visitors or inhabitants of a city to more easily generate a mental image of the city (Darken and Sibert 1993).

To find the legibility elements of a city, we overlay the result of the clusters and groupings over an existing urban dataset that contains information on paths, edges, districts, nodes, and landmarks. By using decreasing levels of detail in the urban model, we are able to correctly identify and rank the elements hierarchically from the most visually important to the least. The ranking allows us to identify the main features that define an urban model globally to minor features that shape the model locally.

Given the ranking of the elements of a city, it is now possible for us to define a city's urban form in a quantitative manner. Figure 9.14 shows the result of applying decreasing levels of detail of the same urban model overlaid on a satellite image of the city. As can be seen, when one uses a coarse level of detail of the urban model, only the most important features of the city are retained, whereas with the finer levels of detail more localized legibility elements become visible.

URBAN MORPHOLOGY — COMPARING TWO DIFFERENT CITIES

Given the ranking of the legibility elements in a city, we are now able to represent and describe a city in a quantitative manner. More important, this quantitative description of a city allows us to compare and identify the differences between different cities in an analytical fashion.

For example, New York City has a gridlike structure, and is generally defined by its boundaries to the river and the ocean. Washington, D.C., has a more radial structure with "rays" (or roads) emanating from the White House and the Congress. In contrast, a growing city such as Charlotte has a more "sprawling" sense to it, as

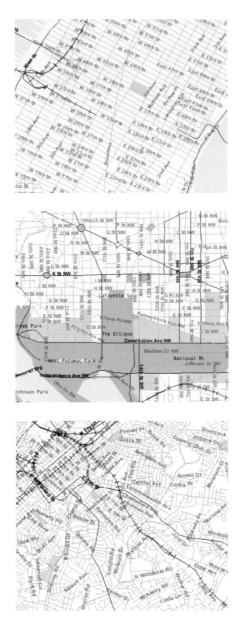

FIGURE 9.15 Views of three cities: New York City (top), Washington, D.C. (middle), and Charlotte (bottom). These three cities have distinctively different layouts; New York City resembles a gridlike structure; Washington, D.C., is radial with roads emanating from the Congress and the White House; and Charlotte is mostly unstructured with strong "sprawling" feeling. **(See color insert after p. 110.)**

new developments are made without the rigorous global planning found in New York or Washington, D.C. (Figure 9.15).

Although most urban planners agree that New York City, Washington, D.C., and Charlotte are very different, they can only describe how these cities are different in a qualitative and subjective way. It has not been possible for urban planners to identify the exact elements that make these cities different or to communicate these differences analytically. With our capabilities, we are starting to categorize cities based on their legibility elements. It may then be possible to start finding quantifiable elements that distinguish American cities from European cities, or older cities from newer cities.

URBAN MORPHOLOGY — HOW A CITY CHANGES OVER TIME

Cities change over time and the legibility elements that define the urban form of the city change over time as well. It is not difficult to imagine a set of legibility elements that distinguishes an urban model in the past, but eventually becomes obsolete as the city grows.

In many small European towns, the towns start with a crossroads, a church, and a town hall, and the intersection of the two roads is the defining "node" of the city. However, as the town grows larger and becomes a city, the church, town hall, and crossroads no longer remain the defining legibility elements for the city. The city of Charlotte is no different. In 1919, when the city's population was less than 60,000, the entire city only occupied what is now the downtown area. At that time, the layout of the city was a relatively structured grid with numbered streets similar to those of New York City (Figure 9.16). As the city grew, however, more roads were built, and

FIGURE 9.16   Map of Charlotte in 1919, when Charlotte was a small city with a clear, grid-like set of roads.

what now defines the urban form of Charlotte is no longer the gridlike structure, but a relatively unstructured network of roads (Figure 9.15).

Using the same framework discussed above for quantifying and ranking elements of urban legibility, we can identify the urban elements from different phases of an urban model over time. We can imagine finding a legibility element that distinguishes an urban model in the past but eventually becomes obsolete as the city grows. Our urban legibility structure will permit us to effectively and compactly organize these elements over time.

VISUALIZING DYNAMIC URBAN ENVIRONMENT

As newer buildings are created in a growing city, older buildings are often destroyed to make space. From a visual and cognitive sense, a single creation or destruction of a building often does not affect the overall legibility of the urban model. This minor visual change, however, needs to be reflected within the underlying hierarchy of the legibility model. In other words, buildings in a city should not be reclustered simply because one single building is created or destroyed. Instead, the newly created or destroyed building should try to obey the existing urban form from a visualization and representation perspective. At some point the legibility model does change enough that reclustering is necessary, however. These may also be significant points in time when the conceptual view of the city changes as well. It could add significant new nodes, for example, or new landmarks and paths. We have devised methods to identify and account for these changes.

ACKNOWLEDGMENTS

This work is supported by the Army Research Office under contract no. W911NF-05-1-0232. We also thank Holly Rushmeier of Yale University and Gabriel Taubin of Brown University for fruitful discussions.

REFERENCES

Bohlen, M.I.H., C. S. Jensen, and M. O. Scholl, eds. 1999. *Spatio-Temporal Database Management: International Workshop STDBM'99.* New York: Springer-Verlag.
Butkiewicz, T., R. Chang, Z. Wartell, and W. Ribarsky. 2007. Analyzing sampled terrain volumetrically with regard to error and geologic variation. Proc. of SPIE Volume 6495, *Visualization and Data Analysis.*
Chang, R., T. Butkiewicz, C. Ziemkiewicz, Z. Wartell, N. Pollard, and W. Ribarsky. 2006. Hierarchical simplification of city models to maintain urban legibility, Charlotte Visualization Center Tech. Rep. CVC UNCC 06–01.
Dalton, R. C. 2002. Is spatial intelligibility critical to the design of large-scale virtual environments? *Journal of Design Computing 4. Special Issue on Designing Virtual Worlds.*
Darken, R. and J. Sibert. 1993. A toolset for navigation in virtual environments. *Symposium on User Interface Software and Technology,* 157–65.
Edelsbrunner, H. 1980. Dynamic data structures for orthogonal intersection queries. Tech. Rep. F59, Inst. Informationsverarb., T.U. Graz, Graz, Austria.
Garland, M. and P. Heckbert. 1997. Surface simplification using quadric error metrics. *Proceedings of SIGGRAPH,* 209–16.

Ingram, R. and S. Benford. 1995. Legibility enhancement for information visualization. *IEEE Conference on Visualization*.

Luebke, D. 2001. A developer's survey of polygonal simplification logarithms. *IEEE Computer Graphics and Applications*, May/June, 24–35.

Lynch, K. 1960. *The Image of the City*. Cambridge, MA: MIT Press.

Maciel, P. and P. Shirley. 1995. Visual navigation of large environments using extended clusters. *Symposium on Interactive 3D Graphics*, 95–102.

Shen, H., L. Chiang, and K. Ma. 1999. A fast volume rendering algorithm for timevarying fields using a time-space partitioning (tsp) tree. *IEEE Visualization'99*, 371–77.

Sillion, F., G. Drettakis, and B. Bodelet. 1997. Efficient impostor manipulation for real-time visualization of urban scenery. *Computer Graphics Forum* 16:207–18.

Voisard, A., P. Rigaux, and M. Scholl. 2002. *Spatial Databases with Applications to GIS*. Morgan Kaufmann, Inc., San Francisco, CA.

Zaniolo, C., S. Ceri, C. Faloutsos, R. T. Snodgrasss, V. S. Subbrahmanian, and R. Zicart. 1997. *Part II Temporal Databases in Advanced Database Systems*. Morgan Kaufmann, Inc.

Zelinka, S. and M. Garland. 2002. Permission grids: practical, error-bounded simplification. *ACM Transactions on Graphics* 21(2):207–29.

10 Mobile, Aware, Intelligent Agents (MAIA)

David A. Bennett and Wenwu Tang

CONTENTS

INTRODUCTION

The movement of energy, matter, and organisms through space and time drives change in geographic domains. The phenomenon of interest can be, for example, inorganic material whose fate and transport are governed by physical law or complex interacting organisms making contextually dependent decisions. In this chapter we discuss ongoing research in GIScience aimed at the digital representation of such objects. Particular attention is paid to the representation and interpretation of mobile, aware, and intelligent agents (MAIA) that interact with one another and the environment within which they live.

It is often difficult to apply traditional scientific methods to the study of dynamic geographic phenomena because: (1) systems often cannot be manipulated experimentally at geographic scales; (2) the temporal lag between cause and effect can be long; and (3) replicates for control are often difficult to establish. Computer simulation of such systems has, therefore, become an important tool in the study of geographic processes, and the scientific community has become increasingly adept at representing certain types of dynamic geographic phenomena. Models that simulate the flow of water across a landscape using deterministic equations of physical laws are, for example, well represented in the literature (e.g., Flanagan et al. 2001, Kang et al. 2006). While modeling overland flow is not a trivial problem, the remaining challenges are largely outside the realm of GIScience. The construction of models that simulate the behavior of mobile, aware, and intelligent organisms that interact with each other and the environment within which they live, on the other hand, presents new and significantly different challenges to the GIScience research community. Although progress has been made in the simulation of such organisms using agent-based modeling techniques, new forms of representation are needed that capture learning, adaptive behavior, spatial and aspatial memory, perception and context, and social as well as geographic topologies. These capabilities produce system level behavior that is far from deterministic and exhibit characteristics commonly associated with complex adaptive systems (CAS) (Holland 1995); characteristics that challenge our ability to create, validate, and interpret models of geographic processes. If these challenges can be adequately addressed, however, the set of problems that could be studied using models driven by geographically savvy digital agents is large:

- Land use and land cover change
- Predator and prey relationships
- Epidemiology and the spread of disease
- Local response to global climate change
- Navigation and wayfinding
- Location-based services
- Diffusion of technology or ideas

OBJECTIVES, CONTEXT, AND CHAPTER ORGANIZATION

Our objectives for this chapter are to discuss the representation, implementation, and interpretation challenges associated with the production of agent-base models of mobile, aware, and intelligent organisms; to describe a general framework for the representation of such agents; and to illustrate the applicability of this framework through a case study of the migratory behavior of elk (*Cervus canadensis*) on Yellowstone's northern range. In section 2 we provide background information and a brief review of relevant literature. A framework for the representation of geographically aware and mobile agents is presented in section 3 and the associated case study is discussed in section 4. In section 5 we conclude with a general discussion of the future of agent-based models of mobile, aware, intelligent entities.

BACKGROUND

Our interests in agent-based models (ABM) of complex adaptive spatial systems lie primarily in the representation of geographically aware, adaptive, and intelligent agents (GAIA). Mobile agents are considered to be an extension to the more general concept of GAIA. More specifically, our research has been focused on how individuals make decisions about: (1) land use and land cover; (2) how to navigate across uncertain and risky landscapes (elk in this situation); and (3) how to organize to effect change in policies that, in turn, effect changes in the production of ecosystem services. Related research is linked via underlying questions about how landscape structure emerges from individual and localized actions and how feedback mechanisms link multiple social or spatiotemporal scales. Agent-based simulation is used as a mechanism to explore the interconnectedness of geographic decision-making and landscape-scale patterns. The design and implementation of associated simulation models draws on four distinct lines of literature: (1) complex adaptive systems; (2) agent-based modeling; (3) machine learning; and (4) spatial modeling. We briefly review the first three issues in the context of spatial modeling.

Complex Adaptive Spatial Systems

Research based on the theory of complex adaptive systems attempts to explain how system level structure emerges from a large number of individual, interacting elements. In geography we can, for example, recast Christaller's central place theory or Von Thunen's theory of land rent as the logical outcome of spatial systems comprised of a large number of identical, fully informed individuals making local decisions about land use on an isotropic plane (Christaller 1966; von Thunen 1966). System-level patterns emerge (e.g., a land rent surface), but do so in a deterministic fashion. If this system is comprised of spatially heterogeneous resources and heterogeneous decision-makers with imperfect knowledge, and the ability to learn from and adapt to changing conditions and to other decision-makers, then it becomes decidedly less predictable and more interesting; the system becomes a complex adaptive spatial system (CASS). Within CASS, opportunity, initiative, and random events can have significant impacts on the path a system takes. An ambitious individual with property at a suboptimal location can, for example, convince the powers that be to build a new road to support his or her development. This action reshapes the "resource surface" that others use to guide future decisions and, thus, produces positive feedback that promotes growth where it otherwise might not have occurred. Furthermore, once a system begins to move down a particular spatiotemporal path it can be costly, or even impossible, to shift it to alternative states (e.g., once development begins around our industrious but suboptimal property owner, it is often impossible to reconstruct the area's original state). Heterogeneity, adaptation, self-organization, and path dependency are characteristics commonly ascribed to complex systems (Cilliers 1998; Levin 1998; Phillips 1999; Manson 2001; Lansing 2003; Reitsma 2003; Carpenter and Brock 2004; O'Sullivan 2004; Manson and O'Sullivan 2006; Portugali 2006; Bennett and McGinnis forthcoming) and apply, we argue, to most spatial systems. A study built on a CASS framework, however, presents significant representation, implementation, and interpretation challenges.

Challenges of Representation

CASS are driven, in part, by the actions of decision makers who are contextually aware, intelligent, and potentially mobile. To simulate the impact that such entities have in dynamic geographic domains, each of these defining characteristics must be considered and represented. Let us begin by considering what is meant by

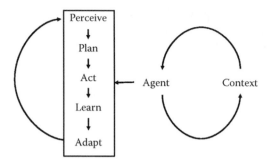

FIGURE 10.1 Agents must adapt to changing context through experiential learning.

"contextually aware." Representing the ever-changing geographic context within which decision makers operate can require a complicated and data-intensive spatio-temporal data structure. At a minimum, agents in an ABM must be able to access these structures and, thus, "perceive" the environment within which they are situated. The contextualization of decision makers can, however, prove even more challenging, because each may have a unique and bounded view of the system within which it operates. Geographic context may be restricted to what individuals see, remember, or learn about and filtered by their own objectives and belief systems. Context can also include the social links that affect an individual's decision-making processes: from whom an individual gets information, whose opinions are respected, who has power over whom, and who interacts with whom. Social context can affect the diffusion of ideas, resources, or disease and establish the economic, political, and social realties within which decisions are made. Finally, if agents are goal-directed they must be cognizant of their own internal context. Relevant context depends on an individual's objectives and could include, for example, measures of heath, wealth, or social status.

Intelligence implies the ability to learn from and adapt to a history of contextualized experiences. More specifically, the designer of intelligent, geographically situated decision-making agents must be cognizant of the sequenced set of steps that transforms perceived stimuli into improved decision-making. These steps are outlined in Figure 10.1. Intelligent agents must perceive real-world context (perhaps imperfectly), construct a plan for responding to that context, act on that plan, learn from the success or failure of that action, and adapt its behavior to reflect what it has learned. New forms of digital representation are needed to adequately reflect decision-making strategies, accommodate spatial and aspatial learning, and store long- and short-term memory.

CHALLENGES OF IMPLEMENTATION

ABM are the current programming paradigm of choice for building intelligent, contextually aware software objects. Systems built using this paradigm have been used to simulate land use and land cover change (Janssen et al. 2000; Parker et al. 2003; Manson 2006; Evans et al. 2006), the routing of pedestrians through buildings and towns (Batty et al. 2003), the spread of disease (Bian 2004; Barrett et al. 2005; Dunham 2005), and animal and habitat relationships (An et al. 2006; Bennett and Tang, 2006; Dumont and Hill 2001; Bian 2001). All of these examples illustrate how agents can be used to simulate geographically situated decision-making, and many model the movement of individual entities through space. Few examples in the geographic literature of ABM, however, are designed to simulate spatial and aspatial learning, or link ABM models to the kinds of spatiotemporal data structures needed to store and query information about dynamic geographic systems. Learning algorithms and spatiotemporal data structures, therefore, present two significant challenges to the implementation of ABM.

Research in the cognitive sciences and machine learning provides insight into how spatial and aspatial learning can be accomplished in ABM of CASS. Cognitive scientists, for example, understand how the brain organizes and stores spatial information (Muller 1996, Shapiro et al. 1997, Eichenbaum et al. 1999) and illustrate how associated biological structures can be mapped into digital domains (e.g., graph-based data structures) (Muller et al. 1996; Trullier and Meyer 2000). Research in machine learning provides examples of how contextualized experience can be mapped to improved decision-making (Sutton and Barto 1998). Progress has been made in GIScience on the implementation of spatiotemporal data structures that capture the movement of digital entities through geographic space (Guting 2005), and work in robotics illustrates how autonomous agents can navigate across a landscape (Arkin 1998). Relatively few studies, however, have tried to tie these various elements of intelligent spatial behavior together in a single implementation (Bennett and Tang 2006).

CHALLENGES OF INTERPRETATION

The goal of complex systems modeling is often to explore system-level behavior as it is produced by a large number of interacting and heterogeneous agents. The ABM produced to simulate CASS are often large and complicated and this by itself poses a significant challenge for the verification, validation, and interpretation of models and model results. While it may seem paradoxical, those models that are best able to simulate the complex behavior of real-world systems are likely to pose the greatest challenge to validation and interpretation. Non-deterministic, nonlinear, path-dependent, self-organized, and emergent behavior are expected from these systems. This suggests that if we accept an explanation based on complexity, then we must also accept the notion that an existing spatial pattern is just one realization of many possible alternative states; non-deterministic systems can produce multiple end states. This potential for multifinality means that a model that fails to reproduce an existing state is not necessarily in error. Similarly, the possibility of equifinality, different processes producing the same end state, suggests that a model that does mimic real-world patterns is not necessarily valid (Figure 10.2, Brown et al. 2006).

Given multifinality and equifinality, how do we prove, for example, that an emergent behavior produced by a simulation is, in fact, generative evidence of real complex behavior, and not an unintended artifact of simplifying assumptions encoded into agent behavior a priori. If generative scientific approaches (Epstein 2007), like ABM, are to be applied to CASS, they must be transparent, but issues of adaptation,

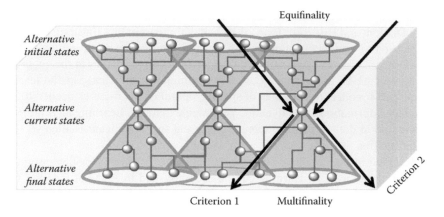

FIGURE 10.2 Interaction, self organization, equifinality, multifinality, and bifurcation are all important in ABM of CASS, but significantly complicate issues of representation, implementation, and interpretation.

equifinality, bifurcation, and divergence, the very same complex behaviors we expect the system to capture, can quickly render the modeling process opaque (Brown et al. 2005). It makes sense to build into complex system models the same kinds of explanatory tools typically associated with expert systems, but tracking cause and effect through a CASS will be considerably more complicated. Can we determine a priori what an important event in an ABM simulation looks like? When, for example, is the variation in the state of some modeled component unimportant noise and when does it signal a bifurcation point? Building into ABM the ability to trace back through model output to gain an understanding of how a system got to where it did is likely to prove challenging, but it seems imperative that we do so if we are to make strong claims about our interpretations of model results.

A FRAMEWORK FOR GEOGRAPHICALLY AWARE INTELLIGENT AGENTS (GAIA)

The GAIA-based simulation framework is specifically designed to support models of CASS that are, in part, driven by the decision-making processes of intelligent, geographically situated entities. It is assumed that spatial decision-making can be a collaborative, multi-objective, and semistructured process supported by limited and uncertain knowledge. Furthermore, it is assumed that decision-makers are able to adapt their spatial behavior to improve performance or respond to changing conditions (i.e., they can learn); without adaptability the system will be "brittle" (Holland 1986) and incapable of representing complex system behavior. In the context of GAIA agents must, therefore, be capable of learning how to: (1) manage spatial resources under uncertainty; (2) organize, compromise, and collaborate to reach individual or societal objectives; and (3) minimize risk and maximize opportunity. Table 10.1 summarizes the characteristics that geographically aware intelligent agents must possess to support these modeling assumptions and objectives. We suggest that these

TABLE 10.1
Characteristics of Geographically Aware Intelligent Agents

Characteristics	Description
Geographic contextual-awareness	Agents must be able to perceive contextual information about themselves and the environment within which they are situated.
Cognitive representation	Agents must maintain an internal representation of their external environment. This representation can be learned though interaction or provided a priori.
Adaptability	Agents must be able to adapt to changing context to achieve goals in complex, risky, and uncertain environments.
Cooperation	Agents must be able to solve problems by cooperating with each other in a direct or indirect manner.
Heterogeneity	Agents can be heterogeneous with respect to properties, knowledge, and behaviors.

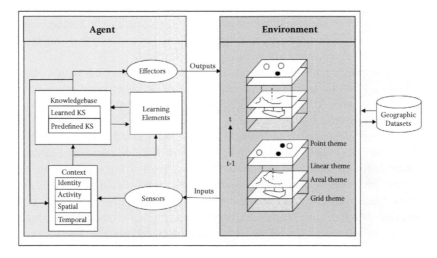

FIGURE 10.3   Conceptual diagram of the *GAIA*-based simulation framework.

characteristics are well suited to the representation of many different kinds of geographically situated decision-making entities.

A GAIA Implementation

In Figure 10.3 we present a generalized schematic of a GAIA framework comprising three primary components: context, knowledge structures, and learning modules. The implementation of this framework is built on the overlapping technologies of agent-based programming, machine learning, and geoinformatics (Figure 10.4). For a more detailed description of the GAIA framework the reader is directed to Tang (2007).

Those elements of the real world that are germane to a simulation are stored as the "environment." This information, along with information about internal and

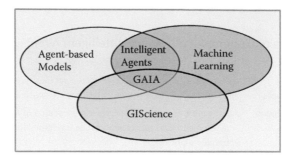

FIGURE 10.4 *GAIA* integrates elements of agent-based modeling, machine learning, and GIScience.

social context, are filtered through an agent's perceptual and cognitive apparatus and stored as part of its internal state. Contextual information is stored in short-term memory and is periodically updated (Smith et al. 1982). Basically, four types of context elements can be maintained in agents situated within spatially explicit environments: identity, activity, spatial, and temporal (Dey et al. 2001). These elements represent primitive situational information for agents and can be further organized into high-level information (e.g., geographic and social context) that supports complex decision-making at individual or collective levels.

The actions of agents (e.g., a decision to do *a* and not *b*) are a function of context. This function can, for example, be a simple probabilistic statement (given context *c*, action *a* will be selected with probability *p*) or a complicated multicriteria evaluation algorithm. Successful action and context couplets are the kinds of objects stored as part of a knowledge structure. These structures, maintained in the long-term memory of agents, guide agents as they make decisions based on contextual information. The knowledge structures of agents can be predefined or learned from agent-agent and agent-environment interactions and are often represented as a set of decision rules modeled in discrete and numeric forms (Russell and Norvig 1995). In GAIA knowledge structures can include spatiotemporal properties modeled as, for example, cognitive maps (Kitchin 1994), which are particularly suitable for characterizing MAIA (discussed in the next section).

The third basic component of GAIA that must be implemented is the ability to learn and adapt. Agents instantiated as part of the GAIA framework should be able to improve their performance (e.g., move toward optimal behavior) and adapt to changing context through direct and indirect experiential learning. As implemented, the learning behavior of GAIA can be modeled at the population and individual levels using evolutionary algorithms, artificial neural networks, or reinforcement learning (Meyer and Guillot 1991; Maes 1994; Bellew and Mitchell 1996; Bennett and Tang 2006).

Mobile, Aware, Intelligent Agents (MAIA)

The ability to move across a landscape is not incorporated into the generalized GAIA framework presented above, because: (1) such behavior is not required for all forms of spatial decision-making (e.g., land-use change); and (2) the modeling of mobility

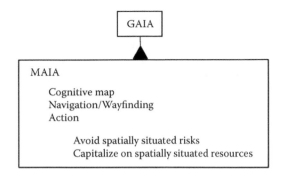

FIGURE 10.5 MAIA is an extension of the GAIA framework.

requires specialized treatment of spatial context, learning, and knowledge. MAIA are, therefore, implemented as a subclass of GAIA (Figure 10.5). More specifically, MAIA are instantiated with: (1) a graph-based cognitive map designed to store learned spatial knowledge of routes; (2) learning algorithms to support navigation and way-finding behavior; and (3) spatial behaviors designed to facilitate resource utilization (geographic regions of attraction) and risk avoidance (geographic regions of repulsion).

MAIA is a work in progress, and fundamental questions remain about how to model spatial learning. In recent work we have examined the effect of alternative learning algorithms on spatial behavior and the resulting success of agents. Two algorithms have been evaluated to date: Hebbian and Q learning. Both algorithms rely on reinforcement mechanisms for problem solving. Hebbian learning is designed to emulate how neurons in the brain might connect actions to positive (or negative) outcomes (Hebb 1949), while Q learning solves problems using temporal-difference learning and delayed-reward mechanisms (Watkins 1989; Sutton and Barto 1998). The literature suggests that both algorithms are biologically feasible and, thus, rational choices for the representation of learning in ABM (Muller et al. 1996; Trullier et al. 1997; Watkins 1989; Sutton and Barto 1998).

AN ILLUSTRATION OF MAIA

The GAIA/MAIA framework was motivated by ongoing research in and around Yellowstone National Park (YNP). The illustration that follows is part of an effort designed to better understand ecosystems dynamics in the area's northern elk winter range (NEWR). The northern boundary of the park bisects the traditional winter range of the northern elk herd and, thus, produces two regions with similar elk habitat but very different management regimes. Over the past 40 years the NEWR has experienced significant change (National Research Council 2002). The park culled the elk population to prescribed levels until the late 1960s. This practice, combined with intensive hunting pressure on non-park land, reduced the northern elk herd to a recorded low of 3,172 in 1968. Few elk ventured outside of the park during this time. In 1968 the park adopted a natural-regulation policy (i.e., the elk population was to be regulated by the carrying capacity of the range) and hunting regulations outside of the park were modified to encourage migratory behavior and increase herd levels

(Lemke 1995). The elk population responded rapidly and grew to a recorded high of 19,000 by the mid 1990s. Elk began to migrate and rediscover winter range outside of the park after the 1988 fire and in 1995 wolves were reintroduced to YNP (Lemke 1998). Since 1995 the elk population has steadily declined (now approximately 9,500) and willow is now growing to heights not seen for 100 years along some of the YNP's riparian corridors (NRC 2002). One explanation for these changes is top-down trophic cascade; wolves reduced the elk population and, as a result, willows were released from intense elk herbivory (Ripple et al. 2001; Fortin et al. 2005). However, other explanations exist. Creel and Winnie (2005), for example, show how wolves can affect the spatial behavior of elk and Vucetich et al. (2005) illustrate how a prolonged regional drought could be responsible for a decline in the elk popula-tion. Finally, Haggerty and Travis (2006) present anecdotal evidence to suggest that changing land use practices outside of the park are affecting the spatial behavior of the region's elk population. Our goal was to develop a modeling framework capable of supporting investigation into landscape-scale dynamics and, in particular, the response of elk to the changing geographic pattern of risks (e.g., wolves and hunters) and resources (e.g., forage, browse, and safe havens). The GAIA/MAIA framework is a result of related efforts.

The MAIA class structure was extended to represent the spatial behavior and bio-energetics of elk. The bioenergetics model used was adapted largely from Turner et al. (1994). Virtual elk were subjected to snow conditions associated with the winter of 1996/97 (a particularly harsh winter). Snow data for this period was spatially distrib-uted using the Yellowstone Snow Model (Farnes et al. 1999). Through repeated inter-action with this context they "learned" when and where to migrate as a function of snow depth, the spatial distribution of forage, and end of winter fitness. For a detailed discussion of this model the reader is referred to Bennett and Tang (2006). Since the publication of these early results we have continued to refine the model and explore the impact of alternative assumptions about spatial learning strategies. The impact of two alternative learning algorithms is evaluated here: Hebbian and Q learning.

RESULTS

To test the impact of alternative algorithms on spatial learning five herds, with 20 elk each, were modeled. The objective of this simulation was to learn migratory routes that maximized end of winter fitness. The initial location of elk is fixed at the cell within the summer range of the Northern Yellowstone herd. Each treatment (Hebbian and Q learning) was simulated 20 times and the simulation was allowed to run through 100 learning episodes (i.e., 100 winters). End of winter elk body mass was used as the mea-sure of learning success and Figure 10.6 shows the learning curves for both treatments. Figure 10.7 illustrates the learned migration patterns for virtual elk using Hebbian learning, Figure 10.8 depicts the migration patterns produced by Q learning.

Differences in the way Hebbian and Q learning algorithms update knowledge structures leads to performance differences. Q learning evaluates the elk's state after each segment in the migratory route is traversed and rewards only that seg-ment (i.e., within-episode learning). With Hebbian learning the entire migratory route is traversed before the state of an elk is evaluated and all segments along that

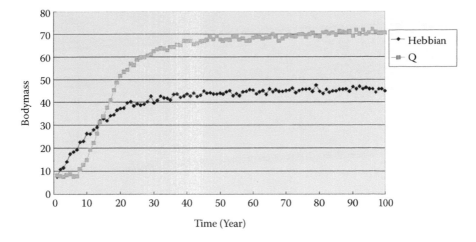

FIGURE 10.6   Learning curves for Hebbian and Q learning.

route receive the same reward (i.e., between-episode learning). As observed in Figure 10.6, Hebbian converges to a maximum value more quickly than Q learning, but the overall performance of Hebbian learning is lower than that of Q learning. The lower performance of Hebbian learning can be attributed to rewarding segments in a migratory route that contribute relatively little to the overall success of that route. By generating more route choices, however, the patterns produced by Hebbian learning facilitate more exploratory behavior than those produced by Q learning. Hebbian learning could, therefore, be more advantageous when conditions along an elk's preferred routes are disturbed (e.g., by fire, wolf predation, or hunting). The possibility

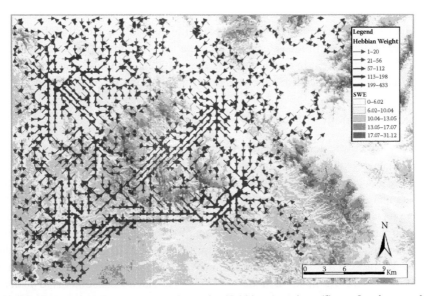

FIGURE 10.7   Migration routes produced by Hebbian learning. (**See color insert after p. 110.**)

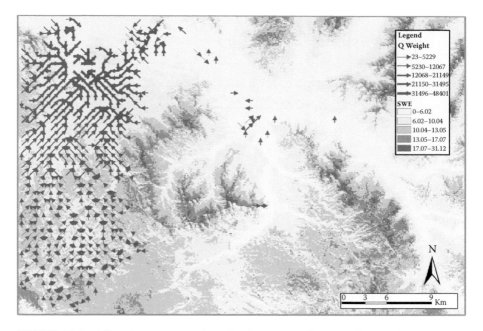

FIGURE 10.8   Migration routes produced by Q learning. (**See color insert after p. 110.**)

that Hebbian-based learning may result in higher elk survival given uncertain and risky environments is being explored.

Reinforcement learning, exemplified by Hebbian and Q learning, employs temporal-difference learning to perform heuristic search for solutions. Optimal solutions cannot be guaranteed using these techniques. Given the limited knowledge and analytical skills of most decision-makers, however, real-world outcomes are not likely to be optimal. Reinforcement learning algorithms can, therefore, be efficient and suitable techniques for the representation of adaptive processes in ABM of CASS. It must be kept in mind, however, that different techniques are likely to lead to differences in convergence speed and, more important, learned outcomes.

CONCLUSIONS

In this chapter we attempted to identify the major challenges associated with modeling dynamic geographic systems — systems that are often driven by the decision-making processes of a large number of mobile, contextually aware, and intelligent agents and exhibit the characteristics of complex adaptive systems. We argue that to capture the complexity of such systems new forms of computer simulation are needed that are capable of representing adaptive spatial learning and the complex set of feedback mechanisms that drive these systems and the individuals who inhabit them. We present a work in progress, the GAIA/MAIA framework, that is designed to move us toward simulation models of CASS, but much work remains to be done in the representation of intelligent, geographically aware entities, the implementation of ABM, and the interpretation of model results.

Given the significant challenges associated with building ABM of CASS, what conclusions can be drawn? A goal of statistically valid models of real-world processes seems a ways off, and perhaps even misguided. Perhaps the greatest value of agent-based models for complex adaptive spatial systems lies in the questions that they require us to ask about system behavior, the way that they require us to conceptualize system structure, and the opportunities that they provide for us to explore plausible outcomes and search for robust decisions.

ACKNOWLEDGMENTS

The authors would like to thank the National Science Foundation (BE-CNH Award #0216588, "Complexity Across Boundaries: Coupled Human and Natural Systems in the Yellowstone Northern Elk Winter Range" and HSD Award #0624297, Collaborative Research: Social Complexity and the Management of the Commons) and The University of Iowa Center for Global and Regional Environmental Research for their support of this project. We also thank our entire research team for their support in research related to this project.

REFERENCES

An, L., G. M. He, Z. Liang, and J. G. Liu. 2006. Impacts of demographic and socioeconomic factors on spatio-temporal dynamics of panda habitat. *Biodiversity and Conservation* 15(8):2343–63.

Arkin, R. C. 1998. *Behavior-based robotics*. Cambridge, MA: MIT Press.

Barrett, C. L., S. G. Eubank, and J. P. Smith. 2005. If smallpox strikes Portland. *Scientific American* 292(3):54–61.

Batty, M., J. Desyllas, and E. Duxbury. 2003. The discrete dynamics of small-scale spatial events: Agent-based models of mobility in carnivals and street parades. *International Journal of Geographical Information Science* 17(7):673–97.

Belew, R. K. and M. Mitchell, eds. 1996. *Adaptive individuals in evolving populations: Models and algorithms*. New York: Addison-Wesley.

Bennett, D. A. and D. McGinnis. Forthcoming. Coupled and complex: Human-environment interaction in the Greater Yellowstone Ecosystem. *Geoforum*.

Bennett, D. A. and W. Tang. 2006. Modelling adaptive, spatially aware, and mobile agents: Elk migration in Yellowstone. *International Journal of Geographical Information Science* 20(9):1039–66.

Bian, L. 2001. The GIS representation of wildlife movements: A framework. In *Remote Sensing and GIS Applications in Biogeography and Ecology*, ed. A. S. Millington, S. J. Walsh, and P. E. Osborne. Boston: Kluwer Academic Publishers.

Bian, L. 2004. A conceptual framework for an individual-based spatially explicit epidemiological model. *Environment and Planning B-Planning & Design* 31(3):381–95.

Brown, D. G., R. Aspinall, and D. A. Bennett. 2006. Landscape models and explanation in landscape ecology — A space for generative landscape science? *The Professional Geographer* 58(4):369–82.

Brown, D. G., S. Page, R. Riolo, M. Zellner, and W. Rand. 2005. Path dependence and the validation of agent-based spatial models of land use. *International Journal of Geographical Information Science* 19(2):153–74.

Carpenter, S. R. and W. A. Brock. 2004. Spatial complexity, resilience, and policy diversity: Fishing on lake-rich landscape. *Ecology and Society* 9 (1). http://www.ecologyandsociety.org/vol9/iss1/art8.

Christaller, W. 1966. *Central Places in Southern Germany*. Englewood Cliffs, NJ: Prentice Hall.

Cilliers, P. 1998. *Complexity and postmodernism: Understanding complex systems*. London: Routledge.

Creel, S. and J. A. Winnie. 2005. Responses of elk herd size to fine-scale spatial and temporal variation in the risk of predation by wolves. *Animal Behaviour* 69:1181–89.

Dey, A. K., G. D. Abowd, and D. Salber. 2001. A conceptual framework and a toolkit for supporting the rapid prototyping of context-aware applications. *Human-Computer Interaction* 16(2–4):97–166.

Dumont, B. and D. R. C. Hill. 2001. Multi-agent simulation of group foraging in sheep: effects of spatial memory, conspecific attraction and plot size. *Ecological Modelling* 141(1–3):201–15.

Dunham, J. B. 2005. An agent-based spatially explicit epidemiological model in MASON. *Journal of Artificial Societies and Social Simulation* 9 (1). http://jasss.soc.surrey.ac.uk/9/1/3.html.

Eichenbaum, H., P. Dudchenko, E. Wood, M. Shapiro, and H. Tanila. 1999. The hippocampus, memory, and place cells: Is it spatial memory or a memory space? *Neuron* 23(2):209–226.

Epstein, J. M. 2007. *Generative social science: Studies in agent-based computational modeling*. Princeton: Princeton University Press.

Evans, T. P., W. J. Sun, and H. Kelley. 2006. Spatially explicit experiments for the exploration of land-use decision-making dynamics. *International Journal of Geographical Information Science* 20(9):1013–37.

Farnes, P. C., C. Htedon, and K. Hansen. 1999. *Snow pack distribution across Yellowstone National Park: Final report*, Department of Earth Sciences, Montana State University, Bozeman, MT.

Flanagan, D. C., J. C. Ascough, M. A. Nearing, and J. M. Laflen. 2001. Chapter 7: The Water Erosion Prediction Project (WEPP) model. In *Landscape erosion and evolution modeling*, ed. R. S. Harmon and W. W. Doe III. Norwell, MA: Kluwer Academic Publishers.

Fortin, D., H. L. Beyer, M. S. Boyce, D. W. Smith, T. Duchesne, and J. S. Mao. 2005. Wolves influence elk movements: Behavior shapes a trophic cascade in Yellowstone National Park. *Ecology and Society* 86:1320–30.

Guting, R. H. 2005. *Moving objects databases*. San Francisco, CA: Morgan Kaufmann.

Haggerty, J. and W. R. Travis. 2006. Out of administrative control: Absentee owners, resident elk and the shifting nature of wildlife management in southwestern Montana. *Geoforum* 37:816–30.

Hebb, D. O. 1949. *The organization of behavior*. New York: John Wiley.

Holland, J. H. 1986. Escaping brittleness: The possibilities of general purpose learning algorithms applied in parallel rule-based systems. In *Machine learning: An artificial intelligence approach II*, ed. R. S. Michaiski, J. G. Carbonell, and T. M. Mitchell. Los Altos, CA: Morgan Kaufmann.

Holland, J. H. 1995. *Hidden order: How adaptation builds complexity*. Cambridge, MA: Perseus Books.

Janssen, M. A., B. H. Walker, J. Langridge, and N. Abel. 2000. An adaptive agent model for analysing co-evolution of management and policies in a complex rangeland system. *Ecological Modelling* 131:249–68.

Kang, M. S., S. W. Park, J. J. Lee, and K. H. Yoo. 2006. Applying SWAT for TMDL programs to a small watershed containing rice paddy fields. *Agricultural water management* 79(1):72–92.

Kitchin, R. M. 1994. Cognitive maps: What are they and why study them? *Journal of Environmental Psychology* 14(1):1–19.

Lansing, J. S. 2003. Complex adaptive systems. *Annual Review of Anthropology* 32(1):183–204.

Lemke, T. O. 1995. A Montana tradition. *Montana Outdoors* March/April: 4–9.

Lemke, T. O. 2005. Personal communication.

Lemke, T. O., J. A. Mack, and D. B. Houston. 1998. Winter range expansion by the northern Yellowstone elk herd. *Intermountain Journal of Sciences* 4(1/2):1–9.

Levin, S. A. 1998. Ecosystems and the biosphere as complex adaptive systems. *Ecosystems* 1:431–36.

Maes, P. 1994. Modeling adaptive autonomous agents. *Artificial Life Journal* 1(1&2):135–62.

Manson, S. M. 2001. Simplifying complexity: a review of complexity theory. *Geoforum* 32(3):405–14.

Manson, S. M. 2003. Epistemological possibilities and imperatives of complexity research: A reply to Reitsma. *Geoforum* 34(1):17–20.

Manson, S. M. 2006. Bounded rationality in agent-based models: experiments with evolutionary programs. *International Journal of Geographical Information Science* 20(9):991–1012.

Manson, S. M. and D. O'Sullivan. 2006. Complexity theory in the study of space and place. *Environment and Planning A* 38:677–92.

Meyer, J. and A. Guillot. 1991. Simulation of adaptive behavior in animats: Review and prospect. From *Animals to Animats: Proceedings of the First International Conference on Simulation of Adaptive Behavior*, Cambridge, MA.

MFWP. 2005. Elk count data provided by T. Lemke, Montana Fish, Wildlife, and Parks.

Muller, R.U. 1996. A quarter of a century of place cells. *Neuron* 17:813–22.

Muller, R. U., M. Stead, and J. Pach. 1996. The hippocampus as a cognitive graph. *Journal of General Physiology* 107(6):663–94.

National Research Council. 2002. *Ecological dynamics on Yellowstone's northern range.* Washington DC: National Academy Press.

O'Sullivan, D. 2004. Complexity science and human geography. *Transactions of the Institute of British Geographers* 29(3):282–95.

Parker, D. C., S. M. Manson, M. A. Janssen, M. J. Hoffmann, and P. Deadman. 2003. Multi-agent systems for the simulation of land-use and land-cover change: A review. *Annals of the Association of American Geographers* 93(2):314–37.

Phillips, J. D. 1999. Divergence, convergence, and self-organization in landscapes. *Annals of the Association of American Geographers* 89(3):466–88.

Portugali, J. 2006. Complexity theory as a link between space and place. *Environment and Planning A* 38(4):647–64.

Pritchard, J. A. 1999. *Preserving Yellowstone's natural conditions: Science and the perception of nature.* Lincoln: University of Nebraska Press.

Reitsma, F. 2003. A response to simplifying complexity. *Geoforum* 34(1):13–16.

Ripple, W. J., E. J. Larsen, R. A. Renkin, and D. W. Smith. 2001. Trophic cascades among wolves, elk and aspen on Yellowstone National Park's northern range. *Biological Conservation* 102(3):227–34.

Russell, S. J. and P. Norvig. 1995. *Artificial Intelligence: A modern approach.* Englewood Cliffs, NJ: Prentice Hall.

Shapiro, M.L., H. Tanila, and H. Eichenbaum. 1997. Cues that hippocampal place cells encode: Dynamic and hierarchical representation of local and distal stimuli. *Hippocampus* 7:624–42.

Smith, T. R., J. Pellegrino, and R. G. Golledge. 1982. Computational process modeling of spatial cognition and behavior. *Geographical Analysis* 14:305–25.

Sutton, R. S. and A. G. Barto. 1998. *Reinforcement learning: An introduction.* Cambridge, MA: MIT Press.

Tang, W. 2007. Development of a spatially-explicit agent-based simulation package for modeling complex adaptive geographic systems. *Proceedings of the UCGIS Summer Assembly*, 2007. http://www.ucgis.org/summer2007/studentpapers/tang_rev060907.pdf.

Trullier, O. and J. Meyer. 2000. Animat navigation using a cognitive graph. *Biological Cybernetics* 83:271–85.

Trullier, O., S. I. Wiener, A. Berthoz, and J. A. Meyer. 1997. Biologically based artificial navigation systems: Review and prospects. *Progress in Neurobiology* 51(5):483–544.

Turner, M. G., Y. A. Wu, L. L. Wallace, W. H. Romme, and A. Brenkert. 1994. Simulating Winter Interactions among Ungulates, Vegetation, and Fire in Northern Yellowstone Park. *Ecological Applications* 4(3):472–86.

von Thunen, J. H. 1966. *Isolated state, an English edition of Der isolierte staat*. Oxford, NY: Pergamon Press.

Vucetich, J. A., D. W. Smith, and D. R. Stahler. 2005. Influence of harvest, climate and wolf predation on Yellowstone elk, 1961–2004. *Oikos* 111(2):259–70.

Watkins, C. J. C. H. 1989. Learning from delayed rewards. PhD Dissertation, Psychology Department, University of Cambridge.

11 Comparing the Growth Dynamics of Real and Virtual Cities

Narushige Shiode and Paul M. Torrens

CONTENTS

INTRODUCTION

Virtual cities were first conceptualized in science fiction literature. They were popularized by cyberpunk authors such as Neal Stephenson and William Gibson and their tales of vast "metaverses" and "cyberspaces," composed of bits rather than atoms, sprawling like megalopolises (Stephenson 1993), towering with information skyscrapers (Gibson 1984), with firewalls crafted to resemble those medieval cities (Gibson 1996). As the stories go, these cyber cities are inhabited by real people, rendered in virtual space as digital avatars of various descriptions while maintaining a corporeal presence in the real, tangible world.

In recent years, the boundary between fact and science fiction has dissolved to some extent. Several such virtual cities have been constructed and continue to evolve online (Activeworlds 2007). The popularity of virtual cities is set to grow still further, catalyzed by the present enthusiasm for Massively Multiplayer Online Role-Playing

Games (MMORPG) and, more generally, for social network services (SNS). Several high-profile virtual cities have been opened, or are imminently forthcoming (Linden Lab 2007; There 2007). In addition, the advent of online 3D city models triggered by the development of various Web-mapping services has increased the opportunity for use of virtual cities for virtual tourism, way-finding, data sharing, and online participatory and decision-making purposes for a variety of applications (Shiode 2001; Shiode and Yin 2008).

Virtual cities have obvious appeal in the context of gaming and entertainment, but they also have other uses that are tied to planning, marketing, designing, promoting, predicting, and assessing and evaluating the various factors surrounding an urban environment and, thereby, forecasting a possible future form of the city (Batty et al. 2001). In particular, researchers in urban studies have long used simulation as a planning support tool (Torrens 2002). The rationale is that simulations can function as virtual laboratories for testing hypotheses, ideas, plans, and policies in artificial urban systems, in ways that are not feasible or may be impractical in the real world. Simulations of this nature generally portray urban systems in very abstract terms — although that is beginning to change; see Benenson and Torrens (2004) for more details on this discussion. In particular, the representation of urban spaces, structures, patterns, and morphologies in traditional urban simulation contexts is often cursory.

Cyber cities that are developed in the MMORPG tradition offer fantastic potential as planning-support tools, because they are populated by real people, replacing the synthetic, "mean individuals" (Wrigley et al. 1996) commonly found in urban-simulation models. The emphasis in classic urban simulation is on mimicking urban processes, generally using a variety of algorithms, equations, or transition rules. In virtual cities, by contrast, urban spaces are configured and generated by real people. Citizens of a virtual city are often collective individual users accessing the virtual environment through their own computer consoles. These virtual citizens, also referred to as netizens, act as designers, architects, planners, developers, policy-makers, and inhabitants in their virtual urban environment (Shiode 2001). Virtual cities thus open up a new world of research potential. Yet little work has been conducted in this area. Research into social interactions, discourse, and behavior in online worlds abounds, and is carried out in fields external to urban studies. Nevertheless, little attention has been paid to virtual cities as urban spaces, despite the obvious advantages of using virtual cities as laboratories for research in urban studies (Shiode 2000). The dynamics of urban space represented by the growth trajectory of virtual cities, in particular, would make a perfect ground for studying the growth and life cycle of an urban environment.

This chapter examines the production of a virtual city — the manner in which urban space is developed and populated — in one of the longest-running virtual cities, AlphaWorld. We are particularly interested in determining the extent to which virtual cities in AlphaWorld resemble their real-world counterparts; in this case, Austin, Texas. We chose Austin, as we identified many common features between the two cities. These include the relatively flat topography, comparable scale of their geographical extent, and rapid growth of the urbanized area over recent years. While the insights obtained through this study may not provide sufficient evidence for the general comparison of both groups of cities, virtual and real, they will help us better understand the prospect of measuring the patterns of urban development in both environments.

Rather than conducting this study from a narrative standpoint, we have applied more rigorous empirical methodologies — from spatial analysis, complexity studies, and GIScience — and we use these to compare and contrast the virtual and the tangible urban environments. Specifically, we employ techniques from a toolkit that we are applying elsewhere to the evaluation of patterns of suburban sprawl in North American cities (Torrens and Alberti 2000) and regularities in the rank-size scaling of metropolitan areas (Batty and Shiode 2003).

The rest of the chapter will be organized as follows. Section 2 reports recent trends in the growth of two case cities, AlphaWorld and Austin, Texas. This is followed by analyses and their results presented in Section 3 with an emphasis on the comparison of the relative complexity and symmetry between the two cities. Section 4 brings the chapter to a close, discussing the findings and their implications in the context of urban studies, and outlining potential avenues for further exploration.

A TALE OF TWO CITIES — ALPHAWORLD AND AUSTIN, TEXAS

AlphaWorld: An Immersive 3D Cyber City

AlphaWorld is an immersive, multi-user, three-dimensional cyber city. Its universe is generated electronically and rendered within the virtual environment of the Internet. Users interface and interact with and within the world by logging in from a networked computer. The world has a surface that extends to an area equivalent to 429,025 sq. km, which is comparable in size to the state of California. AlphaWorld is rendered, in two-dimensional space, as a plane; and its areal extent spreads from a central point, Ground Zero (0N 0W) to four corners — from (32,750S 32,750E) to (32,750N 32,750W). Within this plane, users have constructed three-dimensional cities. The world has been active since 1995. It is still thinly populated, but has a rapid growth rate. As of August 8, 2002, there were 127.6 million objects in AlphaWorld, but only 65.1 million unit cells within its boundary (1.52 percent of the total number of cells) contained manmade objects (Roelofs and van der Meulen 2002).

AlphaWorld is a pseudo-three-dimensional environment in that users can "walk about" in the streets and "talk" to other people. The world is incredibly flexible; its users are free to construct their own buildings and interact with other users (detailed information can be found at www.activeworlds.com). Nevertheless, it also retains certain unique spatial features that are not part of the real world, for example, users can enter arbitrary coordinates and instantly "jump" (or teleport) to the desired destination.

Figure 11.1 illustrates changes in the land cover for AlphaWorld over a five-year period. The maps shown in Figure 11.1 cover the area between (1000N 1000W) and (1000S 1000E) of AlphaWorld, an area equivalent to 400 sq. km (roughly 0.3 percent of the total area of AlphaWorld). Ground Zero is at the center of the picture. Considering the maps in a purely visual sense, a number of striking phenomena are apparent. First, the area has obviously undergone dynamic development. This has been both rapid and complex in its manifestation. In particular, the maps illustrate incremental development in a monotonic and consistent fashion, characteristic of

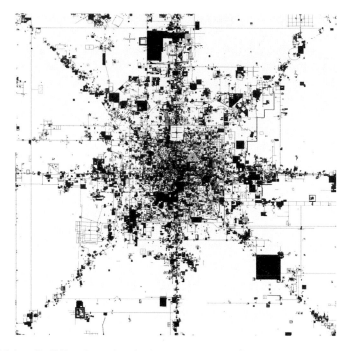

FIGURE 11.1a   Building footprints in the central 400 km² area of AlphaWorld (a) December 1996, (b) February 1998, (c) August 1999, (d) August 2001 (images adapted through a smoothing and reclassification procedure using original images at www.activeworlds.com/community/maps.asp).

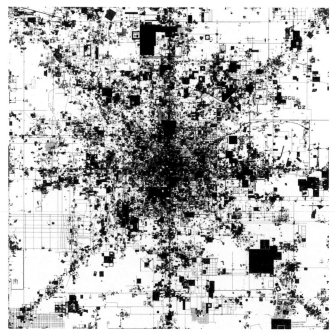

FIGURE 11.1b   (Continued).

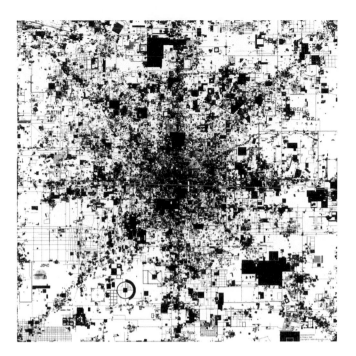

FIGURE 11.1c   (Continued).

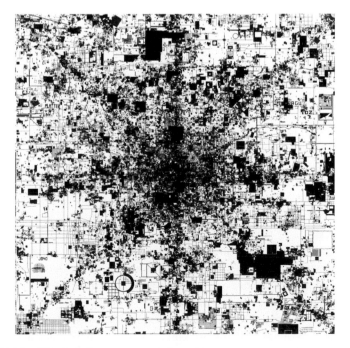

FIGURE 11.1d   (Continued).

TABLE 11.1
The Top Ten Fastest-Growing Cities in the United States

Rank	Metropolitan Area Name	Census Population		Change 1990–2000	
		April 1990	April 2000	Number	%
1	Las Vegas, NV-AZ	852,737	1,563,282	710,545	83.3%
2	Naples, FL	152,099	251,377	99,278	65.3%
3	Yuma, AZ	106,895	160,026	53,131	49.7%
4	McAllen—Edinburg—Mission, TX	383,545	569,463	185,918	48.5%
5	**Austin—San Marcos, TX**	**846,227**	**1,249,763**	**403,536**	**47.7%**
6	Fayetteville—Springdale—Rogers, AR	210,908	311,121	100,213	47.5%
7	Boise City, ID	295,851	432,345	136,494	46.1%
8	Phoenix-Mesa, AZ	2,238,480	3,251,876	1,013,396	45.3%
9	Laredo, TX	133,239	193,117	59,878	44.9%
10	Provo-Orem, UT	263,590	368,536	104,946	39.8%

Source: U.S. Census Bureau, Census 2000 Redistricting Data [P.L. 94-171] Summary File and 1990 Census.

classic monocentric cities. Second, development is focused around a central core area, with a high degree of concentration along two central axes and the diagonals. This "starfish" morphology is due to the unique coordinate teleporting system in AlphaWorld — people tend to focus their building activity along the North-South axis and the "equator," and they also tend to build along coordinates with matching numbers (e.g., 123S 123W). Although these regularities are not at all uncommon in real cities, the degree of concentration is staggering in the case of AlphaWorld. We will revisit this issue later through the comparison of frequency spectra and show that AlphaWorld exhibits by far the more symmetric and regular pattern in its urban growth when compared to Austin.

AUSTIN, TEXAS: A FAST-GROWING REAL CITY

The city of Austin is only the 38th largest city in the United States, occupying 625 sq. km of land in Southern Texas. Nevertheless, it ranks as the fifth-fastest-growing city in the United States (Table 11.1); its population expanded by almost 50 percent in the last decade. Not surprisingly, the urbanized area has expanded at an unprecedented rate (Figure 11.2).

Austin has several similarities with AlphaWorld that make it a good candidate for comparison. Although it was originally founded in the 18th century, Austin's population did not grow above one thousand people until the late 19th century. It has undergone rapid growth and urbanization in very recent years, however, as is the case in AlphaWorld. The land area that Austin occupies (625 sq. km) is comparable to that of the sections of AlphaWorld (400 sq. km) for which we have map data. Incidentally, Texas covers 678,051 sq. km of the United States — roughly 1,085 times larger than

FIGURE 11.2   Urban growth in Austin, Texas, in (a) 1990 and (b) 1995 (growth is shown in gray).

Austin, which is comparable to the entire area of AlphaWorld, 429,025 sq. km, or 1,073 times the size of its central area.

In addition, Austin shares topographical features with AlphaWorld, suitable for developing in all directions. Austin is also a poster child for growing concern about sprawl, urban sustainability, and smart growth; these are some of the issues that we are interested in exploring in virtual cities. In the following section, we will compare the two cities by extracting and examining their spatial patterns. We will also discuss and compare the change in their growth rate to appreciate the magnitude of their sprawling effect.

ANALYSIS

METHODOLOGY

While AlphaWorld exists only within the virtual environment of the Internet, as opposed to its counterpart, Austin, Texas, a real city with a tangible urban structure, we can compare their landscape and growth trajectory by using various imagery-analysis techniques and statistical indexes. We focus on the patterns of, as well as the growth dynamics of, the two cities that are represented by the amount and shape of their urbanized area.

At a glance, both AlphaWorld and Austin show a significant increase in their land-use development over a short period of five years. The rapid growth of their urbanized area and the increasingly complex urban structure are also evident in Figures 11.1 and 11.2. The amount of land-use or the urbanized area in AlphaWorld, in particular, has soared between 1996 and 2001.

The fact that we are only considering their urban form could be somewhat restrictive in that the difference in the amount of the actual socioeconomic activities in each land parcel would not be accounted for. Similarly, the shade of urbanized area reflects on the physical extent of the built environment, and not their residents. Measuring the "population" in a virtual city can be a daunting task, as the users can log in, teleport, and construct buildings wherever they please.

Despite the restriction on the analysis of their socioeconomic and population aspects, interpreting the change in their urban form over time would help maintain an objective perspective for their comparison.

GROWTH ANALYSES

Using the orthographic images from Figures 11.1 and 11.2, the amount of land-use development was measured for each period of the two cities. Both AlphaWorld and Austin share a significant and recent upsurge in urban development over a short period of five years. Their rapid growth and the increasingly complex urban structure are also evident in Figures 11.1 and 11.2. Between 1996 and 2001, land use in AlphaWorld has soared by 130 percent, and now has a high land-coverage rate of 36 percent (Table 11.2). Austin has also seen a significant increase: 27 percent growth in land-use development between 1990 and 1995 (Table 11.3).

In terms of increases in complexity and land coverage in the two cities, we measured their relative transition by examining the changes in their fractal dimensions

TABLE 11.2

Land-Use Development in the Central AlphaWorld

Date	Built Units (% of the area: 1048576)	Standard Deviation	Increase from the Previous Survey
December 1996	164138 (15.38%)	92.66	
February 1998	307502 (29.33%)	116.09	187.34%
August 1999	347794 (33.17%)	120.06	113.10%
August 2001	376700 (35.93%)	122.34	108.27%

TABLE 11.3

Land-Use Development in Austin, Texas

Year	Built Area (% of the area: 1048576)	Standard Deviation	Increase from the Previous Survey
1990	129974 (12.40%)	84.03	
1995	164513 (15.70%)	92.74	126.57%

TABLE 11.4

Change in Fractal Dimensions of AlphaWorld

Date	Fractal Dimension	Correlation Coefficient	Mean Square Error
December 1996	1.729353	0.998244	0.067830
February 1998	1.842562	0.998675	0.062746
August 1999	1.865861	0.998763	0.061389
August 2001	1.880098	0.998855	0.059502

TABLE 11.5

Change in Fractal Dimensions of Austin, Texas

Year	Fractal Dimension	Correlation Coefficient	Mean Square Error
1990	1.664272	0.999663	0.028552
1995	1.701267	0.999608	0.031475

(Tables 11.4 and 11.5). Fractal dimension measures the proportion of a space that is filled and the manner in which it is filled; in general, the higher the value, the denser and more complex the structure (Batty and Longley, 1994). Correlation coefficients are an indication of statistical self-similarity — that is, whether a land-cover pattern retains an overall consistency over a wider area, as well as across different scales.

The values in Tables 11.4 and 11.5 suggest that both cities have undergone consistent growth and have formed an increasingly complex, yet regular pattern within their respective systems. It should be noted that well-established cities like London had a fractal dimension of 1.737 as early as in 1900 and have only observed a moderate increase since then — its fractal dimension in 1962 was still D = 1.774 after 88 years (Abercrombie 1945; Doxiadis 1968). We expect the growth rates of Alpha-World and Austin to drop dramatically within the next decade, as they are quickly approaching the critical limit of space filling: a fractal dimension of D = 2, which would mean a complete filling of the entire plane they occupy. It is popularly known that most spatial structures, including urban systems, have an upper threshold for their growth, which is determined by the amount of space occupied by streets, public parks, and other open area (Batty and Longley 1994).

The fractal dimension of AlphaWorld's central mass (Ground Zero and its environs) is already close to 1.9, which suggests an extremely high density of land use, leaving a small area for streets and open spaces. This is indicative of development patterns to be observed in virtual cities. Because there is no restriction on the form, density, or the structure of their buildings, users of AlphaWorld can afford to build more densely than we would in the real world.

SYMMETRY ANALYSIS

AlphaWorld and Austin also share a common spatial feature in that both of them are highly symmetrical, and this property is evident throughout their growth history. Their similarity ends there, however. The actual shape of land cover in the two cases shows stark contrast, as can be seen in their angular frequency spectra (Figure 11.3). AlphaWorld has a highly diffused spatial structure, with strong regularity around the central axes, whereas Austin has a solid central urban core, surrounded by a less densely inhabited area that sprawls toward every direction in a concentric manner.

Also, if we take the gray level histogram (Table 11.6), AlphaWorld has a higher average value of 91.61 out of a possible 255 (equivalent to 35.93 percent in area coverage) and variance as high as 14,968.05, but its skewness, or an index for the amount of deviation from symmetric histogram distribution:

$$S = \left[\sum_{i=0}^{n-1} (i-\mu)^3 P(i) \right] \Big/ \rho^3$$

TABLE 11.6

Statistical Values of the Gray Level Histogram

Index	AlphaWorld	Austin, TX
Mean	**91.61 (35.93%)**	31.61 (12.40%)
Variance	**14968.05**	7060.97
Skewness	0.59	**2.28**
Kurtosis	1.34	**6.21**

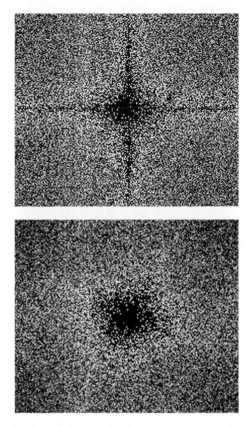

FIGURE 11.3   Visualization of the angular frequency spectrum ($-\pi \sim \pi$) in (a) the central 400 km² area of AlphaWorld as of August 2001 and (b) the central area of Austin, Texas, as of 1995.

as well as kurtosis, an index for the degree of concentration of its histogram distribution around the average value:

$$K = \left[\sum_{i=0}^{n-1} (i - \mu)^4 P(i) \right] \bigg/ \rho^4$$

are found to be at a lower level of 0.59 and 1.34, respectively. By contrast, the gray level histogram for Austin in 1990 has a lower average value (E = 31.61 out of a possible 255, which is equivalent to 12.40 percent in area coverage) and lower variance (V = 7060.97), but much higher skewness (S = 2.28) and kurtosis (K = 6.21). These values confirm the contrast between the urban concentrations found in the two cities, where AlphaWorld has a highly regular structure with much less deviation in its granularity, whereas Austin shows an overall concentric structure but with a varying degree of density.

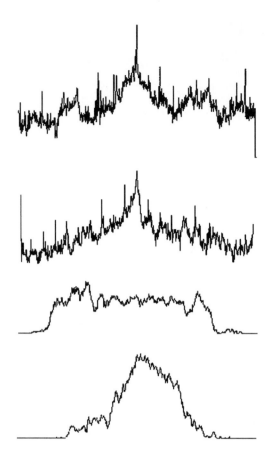

FIGURE 11.4   Projections of density distributions in central AlphaWorld (as of 2001) and Austin (as of 1995): (a) horizontal projection in AlphaWorld, (b) horizontal projection in AlphaWorld, (c) horizontal projection in Austin, and (d) vertical projection in Austin.

We also performed vertical and horizontal projections of density distribution of the two cities (Figure 11.4). Density distribution is a widely used notion in molecular physics and biophysics, but its concept is also applicable to other fields such as the measurement of urban structure. In this case, we are measuring the projection of the cumulative density of built environment along the north-south axis and the east-west axis in the two cities. While they both maintain an overall symmetry in their structure, AlphaWorld is much more dispersed over its landscape than Austin, mainly because of the concentration around the outreaching axes and diagonal lines (Figures 11.4a, 11.4b); whereas in the case of Austin, there is a significant concentration toward the middle section of the map, that is, its downtown (Figures 11.4c, 11.4d).

On the other hand, Figure 11.4 also shows the relative complexity in the urban landscape of AlphaWorld where its frequency fluctuates at a much shorter interval. This same pattern is evident in the statistical texture description. The energy level

$$\left(\sum_{i=0}^{n-1}\sum_{j=0}^{n-1}(i,j)^2\right)$$

across the density distribution is consistent — 0.76 ($\theta = 0$, $\pi/4$, $\pi/2$, $3\pi/4$) for Austin (as of 1995); AlphaWorld has a much lower range at 0.40 ($\theta = 0$, $\pi/2$) and 0.36 in diagonal directions, which again confirms the overall diffusion. In addition, the inertia levels (i.e., the resistance to changes in density) for AlphaWorld range from 151.47 ($\theta = 0$) to 221.82 ($\theta = 3\pi/4$), which are a magnitude larger than those of Austin: 16.89 ($\theta = 0$) to 21.70 ($\theta = 3\pi/4$). All these statistics suggest that, while both AlphaWorld and Austin have highly symmetrical land cover patterns, Austin has a strong concentric structure, as compared to the radial pattern of AlphaWorld.

IMPLICATIONS OF THESE FINDINGS

LOCATION, LOCATION, LOCATION!

The results from the two separate analyses suggest that AlphaWorld and Austin have a similar growth trajectory yet very different morphological geography. That both cities are booming with rapid increase in their urbanized area gives an impression that their patterns of urbanization are similar in nature.

However, the results from the angular spectrum analyses clearly show that the urban development that is taking place in AlphaWorld is by far the more regular and symmetrical phenomenon than that in Austin. Its users exhibit a strong preference toward locations along the vertical, horizontal, and diagonal axes that radiate from Ground Zero, or the central node. As aforementioned, it reflects the unique address system and preferred way-finding heuristic used in AlphaWorld, where each location is identified by a standard Cartesian coordinate system. As users can teleport to any location with a click of a button, "addresses" with specific combination of coordinate values become a valuable commodity and a new prime location that is desirable in the virtual market. Here in the virtual city, accessibility in the traditional sense of having shorter Euclidian distance gives way to the significance expressed by a sequence of coordinates.

In contrast, the omni-directional pattern of urban sprawl observed in Austin suggests the persistence of accessibility in the traditional sense that assumes a monotonic distance-decay function of land price originating in the town center. It is indicative of our choice of location in a real city that emphasizes better access to a strong, growing CBD. At the same time, constant expansion into the suburbs and exurbs implies sprawl on Austin's fringes. The voracious expansion of suburban area is indicative of a phase-shift in preferences for such locations.

That the two cities share a similar growth trajectory but exhibit a different growth pattern shows that both worlds are equally attractive to the residents but that there is a difference in the preferred "location" between the two worlds. After all, patterns of urban growth found in both virtual and real cities reflect a transition in the desired and prime locations that are sought after by their residents.

BUILDING THE IMPOSSIBLE

The other marked difference between the two cities was found in their vertical profile, or the projections of density distributions (Figure 11.4). For one thing, the frequency range for the density distribution of AlphaWorld is much wider. With the ability to teleport themselves to any location of their choice, users of AlphaWorld have access to geographically farther, remote locations and can, therefore, afford to sprawl outward. This is one luxury we do not have in the real world, as remote locations usually mean less accessibility or opportunity. The relatively confined density distribution of Austin shows that the conventional geography is still intact in this real city.

Besides this, we can also appreciate the difference in the complexity of the wave form of the distribution projections. The dataset from AlphaWorld shows a much denser and vibrant fluctuation in its projection. There are a couple of reasons for this:

1. Many of the buildings designed by the users in a virtual city take an elaborate and intricate form that would be difficult, both technically and financially, to reproduce in the real world.
2. The density and the capacity of each street in a virtual city need not be controlled, as there is no real traffic or need for setback of housing facades to be instituted, and users tend to build much more densely than they would in the real world.
3. In short, buildings in a virtual city tend to adopt different forms and have a larger mass than in the real world. It is likely to remain that way unless a building code or a policy to restrict such building behavior were introduced, although we would not have the necessity to do so in the virtual world.

CONCLUSIONS

This chapter presented results from a comparison of urbanization trends in two fast-growing cities, one in real and the other in virtual space: Austin, Texas, and Alpha-World. Preliminary exploration shows some striking similarities between the two cities. Both are sprawling at unprecedented rates and in similar styles. The rate of growth is also comparable, as is the area of land cover. Both are developing at lower-than-average densities on the periphery of the urbanized mass, both have strong central cores amid peripheral expansion, both exhibit symmetrical structures, and both have fractal signatures that indicate a high degree of scattering and fragmentation in land use.

Nevertheless, there are important differences. While both cities are sprawling in a similar fashion, the nature of that sprawl varies considerably between the two cases. Closer analysis reveals significant variation in the pattern of urbanization. AlphaWorld has strong axial orientation, radiating from its central area in horizontal, vertical, and diagonal vertices. Austin, on the other hand, has strong concentric clustering in its spatial distribution. Spectral analysis demonstrates the statistical significance of the variation.

Austin appears to follow the classic stereotype for suburban sprawl in North American cities: a loose and fragmented suburban ring orbiting a central core. Alpha-World, on the other hand, appears to exhibit two urbanization trends simultaneously: classic sprawl, but organized in a polycentric fashion. The end result, in the case of AlphaWorld is a souplike constellation of fragmented urban clusters.

Of course, we are discussing a virtual city, but the implications for urban planning and policy in the real world could be profound. To a certain degree, the construction of urban space in AlphaWorld is indicative of the types of urban forms that people would like to build in the real world. Our analysis has demonstrated that the pattern of urbanization in AlphaWorld is a hybridization of two very popular trends in Western development and settlement patterns. The behavior of people in this cyber city may seem like science fiction: people building without concern for zoning regimes or planning codes, with little care for distance and the capacity to move relatively ubiquitously across the landscape. This is far from far-fetched in real world contexts, however. The situation in AlphaWorld mirrors current trends in the real world: development in the southwest of the United States, for example, appears in many cases to be exercised with little consideration for planning. Decentralized and unplanned growth is the norm, also, on the outskirts of many cities in Latin America. The ubiquity of highways and the public's stated preference for vehicular travel is also creating environments in which space is beginning to matter less and less (Giuliano 1989). To a certain degree, then, AlphaWorld is suggestive of future trends in urbanization and is perhaps the perfect laboratory for exploring what-if questions in urban studies.

REFERENCES

Abercrombie, P. 1945. *Greater London Plan 1944*. London: HMSO.

Activeworlds. 2007. *AlphaWorld 3.3*. Activeworlds Corporation, Newburyport, MA, www.activeworlds.com.

Activeworlds. 2008. Activeworlds Maps. http://www.activeworlds.com/community/maps.asp.

Batty, M., D. Chapman, S. Evans, M. Haklay, S. Kueppers, N. Shiode, A. Smith, and P. M. Torrens. 2001. Visualizing the city: Communicating urban design to planners and decision-makers. In *Planning Support Systems*, ed. R. Brail and R. Klosterman, 405–43. Redlands, CA: ESRI Press.

Batty, M. and P. Longley. 1994. *Fractal Cities*. London: Academic Press.

Batty, M. and N. Shiode. 2003. Population growth dynamics in cities, countries and communication systems. In *Advanced Spatial Analysis*, ed. P. Longley and M. Batty,, 327–44. Redlands, CA: ESRI Press.

Benenson, I. and P. M. Torrens. 2004. *Geosimulation: Automata-Based Modeling of Urban Phenomena*. London: John Wiley & Sons.

Dodge, M. 2000. *Mapping a virtual city, Mappa Mundi*. http:/www.mundi.net/maps/maps_013/.

Doxiadis, C. A. 1968. *Ekistics: An Introduction to the Science of Human Settlements*. London: Hutchinson.

Gibson, W. 1984. *Neuromancer*. New York: Ace Books.

Gibson, W. 1996. *Idoru*. London: Penguin Books.

Giuliano, G. 1989. New directions for understanding transportation and land use. *Environment and Planning A* 21:145–59.

Linden Lab. 2007. *Second Life*, Linden Research, Inc., San Francisco, http://secondlife.com/.

Shiode, N. 2000. Urban planning, information technology, and Cyberspace. *Journal of Urban Technology* 7(2):105–26.

Shiode, N. 2001. 3D urban models: Recent developments in the digital modeling of urban environments in three-dimensions. *GeoJournal* 52(3):263–69.

Shiode, N. and L. Yin. 2008. In press. Spatial-temporal visualization of built environments. In *Understanding Dynamics of Geographic Domains*, ed. K. Stewart Hornsby and M. Yuan. Boca Raton: CRC Press.

Stephenson, N. 1993. *Snow Crash*. New York: Bantam Books.

There (Beta) 2007. *There*, Menlo Park, CA, www.there.com.

Torrens, P. M. 2002. Cellular automata and multi-agent systems as planning support tools. In *Planning Support Systems in Practice*, ed. S. Geertman and J. Stillwell, 205–22. London: Springer-Verlag.

Torrens, P. M. and M. Alberti. 2000. *Measuring sprawl*. London: University College London, Centre for Advanced Spatial Analysis.

Wrigley, N., T. Holt, D. Steel, and M. Tranmer. 1996. Analysing, modelling, and resolving the ecological fallacy. In *Spatial Analysis: Modelling in a GIS Environment*, ed. P. A. Longley and M. Batty. Cambridge: GeoInformation International.

Index

A

Accuracy
  gazetteers, 53, 62, 63
  geocoders, 64, 66, 68
  mediators for gazetteer production, 60
  volumetric terrain analysis, balancing
    accuracy and computation time, 158,
    206
Activity-travel behavior, 96, 97, 98, 100, 107,
    108, 110
Acyclic network properties, 35
ADL Gazetteer, 53, 54
Aerial photographs, 136, 137–138
Aesthetic models, 138
Agent Analyst, ArcGIS, 17
Agent-based models (ABM)
  domain of geographic dynamics, 15
  MAIA (mobile, aware, intelligent agents),
    171–183; see also MAIA (mobile,
    aware, intelligent agents)
Aggregation of models, 162, 163, 207
Airborne oblique photography, 151
Airborne survey data, 136
Aircraft track modeling, 19
Air pollution, 80, 81, 87
Air-pollution dispersal models, 80
Air transportation, 37–41, 43–44
Alexandria Digital Library (ADL) gazetteer, 53,
    54, 55
Algebra, relational, 23
Algorithms, GIScience role, 16
AlphaWorld, 140, 187–201
Anecdotal knowledge, 16
Angular frequency spectrum, 197
Animations, 3, 10, 79
Anomalies, human space-time behavior, 94
Anomalous centrality, 43
Aquarium, space-time, 98–99
ArcGIS
  3D Analyst, human space-time behavior, 103,
    104
  Agent Analyst, 17
  GIScience tools, 17, 19
  representations of flows, 19, 20–21
Architectural details, 136
ARC/INFO GIS, 107, 120
ARC Macro Language (AML), 107
ArcView GIS, 110
Areal measures, 117

Arsenic in drinking water, 78, 79–80, 82, 83–85,
    86, 87
Artifacts, spectral difference image, 123
Artificial intelligence, 59–60, 68
Atmospheric effects, 123, 125
Atmospheric science, modeling in, 14
Atomic level, gazetteer information, 53
Austin, Texas, 187–201
AutoCAD, 142
Autocorrelation, 116, 117, 119, 123
Automata, 17, 23
Automation/automated methods
  gazetteer production, 54, 55–58, 62
  search techniques, 137
  spatial metrics, 115, 116, 120, 129
  urban model reconstruction, 151
Autonomous agents, 17
Avenue, ArcView GIS, 110
Average value, real versus virtual city gray-level
    histogram, 197
Awareness, contextual, 173–174

B

Behavioral pattern recognition, humpback
    whales, 7–10
Bifurcation, 176
Block extrusion, 136
Block models, 135, 136
Bridging, networks, 44
Buffalo, New York, 141–146
Building detail, three-dimensional urban model
    trends, 137–138, 139
Built environments, see Urban growth dynamics,
    real versus virtual cities; Urban
    models, three-dimensional; Urban
    terrain dynamics

C

C (language), 23
CAD models, 136, 137, 138–139, 204
Camera images, relating visual changes to spatial
    metrics, 120–125
Cartography, 14
CASE tool, 18
Categorical maps, 118
Cellular automaton, 14, 17, 23

V

T - #0167 - 171019 - C8 - 234/156/11 - PB - 9780367387525